Telecommunications Policy-Making in the European Union

Please note that the views set forth in this book are the personal views of the author and do not necessarily reflect those of the law firm Skadden, Arps, Slate, Meagher & Flom, LLP.

Telecommunications Policy-Making in the European Union

Joseph W. Goodman

Visiting Scholar, BMW Center for German and European Studies, Georgetown University and Associate, Skadden, Arps, Slate, Meagher & Flom, LLP, USA

Edward Elgar

Cheltenham, UK • Northampton, MA, USA

© Joseph W. Goodman, 2006

Published by
Edward Elgar Publishing Limited
Glensanda House
Montpellier Parade
Cheltenham
Glos GL50 1UA
UK

Edward Elgar Publishing, Inc.
136 West Street
Suite 202
Northampton
Massachusetts 01060
USA

A catalogue record for this book
is available from the British Library

Library of Congress Cataloguing in Publication Data
Goodman, Joseph W. (Joseph William)
    Telecommunications policy-making in the European Union / Joseph W. Goodman.
        p. cm.
    Includes bibliographical references and index.
    1. Telecommunications policy—European Union countries. 2. Telecommunication—European Union countries. I. Title.

    HE8085.G66 2006
    384.094—dc22

                                                                2005052822

ISBN-13:   978 1 84376 806 7
ISBN-10:   1 84376 806 2

Printed and bound in Great Britain by MPG Books Ltd, Bodmin, Cornwall

This book is dedicated to the memories of my brother Bob and sister Kathy
– their spirits will live on in my heart and soul forever

Contents

PART I   THEORETICAL AND HISTORICAL BACKGROUND

PART II   THE HARMONISATION OF EUROPEAN
          TELECOMMUNICATIONS POLICIES

PART III   THE LIBERALISATION OF EUROPEAN
           TELECOMMUNICATIONS

PART IV   TOWARDS A NEW REGULATORY FRAMEWORK
          AND THEORETICAL APPROACH

Figures

Tables

Abbreviations

APEC	Association of Private European Cable Operators
ART	*Autorité de Régulation des Télécommunications*
BEUC	*Bureau Européen des Unions de Consommateurs*
BT	British Telecom
CEC	Commission of the European Communities
CEPT	*Conférénce Européene des Administrations des Postes et des Télécommunications*
CGT	*Confédération Générale du Travail*
COM	Council of Ministers
COREPER	Committee of Permanent Representatives
CPE	Customer Premise Equipment
CTC	Community Telecommunications Committee
DBP	Deutsche Bundespost
DG III	Internal Market Directorate-General
DG IV	Competition Directorate-General
DG XIII	Telecommunications Directorate-General
DGT	*Direction Général de Telecommunications*
DPG	DBP Workers Union
DTI	Department of Trade and Industry (British)
EAIS	European Association of Information Services
ECC	Electronic Communications Committee
ECJ	European Court of Justice
ECTEL	European Conference of Telecommunications and Electronics Industries
ECTRA	European Committee on Telecommunications Regulatory Affairs
ECTUA	European Council of Telecommunications Users' Associations
ECU	European Currency Unit
EMAC	Economic and Monetary Affairs Committee (European Parliament)
EMU	European Monetary Union
ERA	European Regulatory Authority
ERC	European Radiocommunications Committee

ERO	European Radiocommunications Office
ERT	European Round Table of Industrialists
ETNO	European Association of Telecommunications Network Operators
ETO	European Telecommunications Office
ETP	European Telecommunications Platform
ETSI	European Telecommunications Standards Institute
FO	*Force Ouvrière PTT*
GAP	*Group d'Analyse et de Prévision*
IBC	Integrated Broadband Communications
ICC	International Chamber of Commerce
INTUG	International Telecommunications Users' Group
ISDN	Integrated Services Digital Network
ITTF	Information Technologies Task Force
MCR	Merger Control Regulation
MEP	Member of the European Parliament
NERA	National Economic Research Associates
NRA	National Regulatory Authority
OFTEL	Office of Telecommunications (British)
ONP	Open Network Provision
ONP-CCP	ONP Consultation and Coordination Platform
PTT	Post, Telephone and Telegraph Administration
QMV	Qualified Majority Voting
RACE	R&D in Advanced Communications Technologies in Europe
SEA	Single European Act
SEM	Single European Market
SOGT	Senior Officials Group on Telecommunications
TEU	Treaty on the European Union
TO	Telecommunications Operator
UNICE	Union of Industrial and Employers' Confederations of Europe
VANS	Value Added Network Services

Foreword

After years of progressive liberalisation that ended in 1998, the European Union has recently entered a new era in European telecommunications regulation, as Joseph Goodman clearly explains. This book is a welcome commentary on and offers insights into the relevant history, legislation and literature of European telecommunications liberalisation and harmonisation. I am sure that it will be a valuable resource for academics and professionals, as well as for all interested in the regulatory framework for electronic communications in the European Union.

Compared with the former regulatory framework that included 22 EU Directives, the main legislative instruments of the new framework comprise only six Directives. The new framework modernises and updates the old rules in new, technology neutral ways that reflect the emerging multi-platform convergent environment. This new framework is a blueprint for light touch rules to encourage market entry and to regulate network operators and service providers only when markets fail.

As Joseph amply demonstrates, the old Community framework launched a progressive and meaningful liberalisation of European telecommunications markets. This delivered higher quality services and lower prices for consumers. However, technology and markets have evolved, and the rules needed to be modernised to deal with today's market landscape of converging networks and services. Building on a technology neutral approach, the new regulatory framework reflects trends in convergence and will create a stable and predictable regulatory environment, encourage innovation and stimulate new investment in communications networks and services, by both new entrants and existing operators. Broadband technologies will be rolled out quickly and will make a substantial contribution both to competition and to the economic recovery of the sector.

Achieving better regulation is a common challenge for the 25 Member States and the EU institutions, as well as for businesses and consumers. It means finding a balance between the protection of the public interest and the burden of regulations, which may damage the prospects for competitiveness, sustainable growth, employment and trade. The intention is both to encourage efficient investment in infrastructure and also to promote innovation. The objectives which are inbuilt into the new framework are to promote competition, to protect the citizen and to consolidate the internal market. This will help to guarantee the long-term sustainability of competitive markets to the benefit of all.

I am convinced that the European Union now has a modern and predictable regulatory environment for electronic communications, which strikes a fine balance between flexibility and competition necessary for a growing sector. It is well suited to attain the longer-term aim of more sustainable facilities and to achieve a situation where there is competition between different services and infrastructures.

Sandra Keegan
European Commission, Directorate-General for the Information Society
January 2005

Introduction

This book is about the emergence of a European Union (EU) telecommunications policy.[1] The foundation of the EU's policy was laid in 1987, when the European Commission published its seminal Green Paper on Telecommunications. The Green Paper put forth a two-pronged strategy of liberalisation and harmonisation, which set the future course of policy development in the sector. This book concentrates on the Commission's two-pronged strategy, which culminated in an important milestone on 1 January 1998, when the EU Member States fully opened their telecommunications markets to competition. Another milestone was achieved on 1 July 2003, when the EU Member States implemented one of the most sophisticated regulatory frameworks in the world, which is adapted to the liberalised environment and the convergence of electronic communications networks and services.

Political scientists dispute whether the Commission has led the Member States throughout this process (Sandholtz 1998; Sandholtz and Sweet 1998) or partnered and cooperated with them (Thatcher 2001). In fact, both tendencies have been evident. Indeed, beginning in the late 1970s, the Commission has served as a policy entrepreneur in telecommunications. It institutionalised the consultation of affected interests at the EC level; marshalled political consensus in favour of reform; and coordinated the EC-wide liberalisation of the sector. At the same time, however, policy development has been highly sensitive to national preferences. Policy output on important issues has generally reflected the preferences of the UK, French and German governments and, in key areas, the Member States have been able to resist the transfer of authority to the EU.

This book is the product of six years of research as a doctoral student at Oxford University. Eighteen months were spent in Brussels, in and around the Community institutions and Brussels-based organisations, where I collected archive materials and conducted more than 150 interviews with European policy makers. Interviews were also conducted with national policy makers in three national capitals. The panel included representatives from: the European Parliament (EP), and within the EP, the Economic and Monetary Affairs Committee (EMAC); the European Commission, and within the Commission, the Information Society Directorate-General (formerly DG XIII) and the Competition Directorate-General (formerly DG IV) and the cabinets of the Competition and Telecommunications Commissioners; the Council of Ministers, including national representatives; the Economic and Social Committee; the Committee of the Regions; the *Conférence Européenne des Administrations des Postes et des Télécommunications*; user interest groups; national regulators and government officials; telecommunications operators and independent consulting firms.

In the process of researching and writing this book, thanks are due to many. First and foremost, the completion of this book would not have been possible without the support of my parents, Allan and Barbara Goodman. They did more than they realise to ensure its success. The Weiner-Anspach foundation funded a profitable year at the *Université Libre de Bruxelles*. During this time, Kjell Eliassen provided me with useful information and advice, particularly with regard to collecting data and conducting interviews in Brussels. Mark Thatcher, at the London School of Economics, gave me helpful advice on telecommunications issues towards the beginning of this project, which helped to get my research off in the right direction. John Peterson, at the University of Glasgow, gave a number of important criticisms in my early attempts at synthesising a theoretical framework. David Goldey, at Lincoln College, Oxford University, also provided me with helpful advice in revising my theoretical approach. The BMW Center for German and European Studies at Georgetown University and the Washington DC office of Skadden, Arps, Slate, Meagher & Flom, LLP provided valuable support and assistance in the final stages of this book

At Oxford University, I was supervised by Sonia Mazey. Sonia gave unceasingly in her time, assistance and expertise and I owe her an enormous debt. My D. Phil examiners, David Hine and Eleanor Ritchie, offered valuable criticism as well as encouragement. Finally, I am eternally grateful to Vincent Wright. This project would never have gotten underway without his support. As my initial supervisor at Oxford, he shared with me a wealth of knowledge and insight into European politics and helped to narrow the focus of my research – his enthusiasm for the subject matter inspired me to finish this project. I learned a tremendous amount from him and am extremely fortunate to have had the opportunity to work with him.

Despite these acknowledgements, I take full responsibility for the contents of, and any errors contained within, the book. The book puts forward a series of 'snapshots' of the telecommunications policy-making process in Europe from 1957 to 2003 and is divided into four Parts. Part I puts forth the theoretical approaches relied on in the book (chapter one); the historical background of telecommunications policy development in the Member States (chapter two) and early Commission involvement in the sector (chapter three). Parts II and III examine the Commission's two-pronged strategy of liberalisation and harmonisation, as first set forth in its 1987 Green Paper on Telecommunications. Different institutional settings are analysed, including the EU's normal legislative procedures used to re-regulate or harmonise telecommunications legislation (Part II), and the EU's competition rules used to liberalise the telecommunications sector (Part III). Part IV discusses the development of a new regulatory framework, the 2003 electronic communications framework (chapter nine), and then concludes with a discussion of the wider implications of the findings of the book and how the book contributes to the development of a theoretical approach for EU policy making (chapter ten).

This book is directed to those actively involved in, and those interested in learning about, the regulatory environment of the European telecommunications sector. It is also directed to students of European integration who would like to understand how policy has developed in one of the most important sectors in the global economy. New institutionalism is argued to be a useful starting point for analysing the EU policy-making process, but it is further argued that, on its own, it does not adequately account for how policy developed in the sector. In applying multiple policy-making theories to an in-depth analysis of a single sector, this book is meant to contribute to the discussion of the best way to analyse European policy making. In this regard, this book hopes to shed light on the utility of a synthesis of theoretical approaches, applied to various stages, or levels, of analysis, for understanding and explaining the many dimensions of EU policy making.

NOTES

1. Because the analytical focus of the book is on European Community (EC) policy making, the book refers to the EC interchangeably with the EU.

PART I

Theoretical and historical background

Theoretical and Historical Background

1. Conceptualising policy making in European telecommunications

INTRODUCTION

The European Commission was not provided with explicit authority to propose telecommunications legislation in the 1957 Treaty of Rome. Historically, the sector was viewed as a national concern and characterised by a strong public service monopoly tradition. The Member States sought to preserve their sovereignty over policy development and resisted European Commission encroachment over their work. Beginning in the 1980s, however, the sector was progressively liberalised and EU competence in the sector gradually expanded. Even in the absence of an explicit Treaty base legitimising Commission policy leadership, a comprehensive EU regulatory framework was developed and largely implemented by the end of the twentieth century.

This book explains how and why the EU's telecommunications policy developed and why certain reforms in the sector were easier to achieve than others. It is, in other words, a detailed analysis of the EU telecommunications policy-making process. The analytical framework employed to explain the policy changes outlined draws upon new institutionalism and policy actor based approaches. Although new institutionalism is a useful beginning point for analysis, it cannot, by itself, adequately account for how policy developed in the sector. Thus, in addition, the book puts forth a synthetic approach for examining and explaining EU policy making.

1.1 NEW INSTITUTIONALISM: A STARTING POINT FOR ANALYSIS

New institutionalism provides a useful analytical approach for understanding the EU policy-making environment; institutions structure the access of socioeconomic and political actors to the policy-making process and structure debate. Institutions, therefore, play a key role in determining policy outcomes. This approach also helps to explain the development of the policy

3

environment at the European level. Supranational institutions, especially the European Commission, with the support of the European Court of Justice (ECJ), played a key role in advancing ideas, mobilising interests and promoting supranational solutions to the common challenges faced by the Member States. Two variants of new institutionalism are drawn upon: historical institutionalism and rational choice institutionalism.[1]

Historical institutionalism is employed to trace the overall development of institutions and policies in the European telecommunications sector from 1957 until 2003. One of the earliest case studies to rely on historical institutionalism is Peter Hall's *Governing the Economy* (1986). Hall (p. 19) contends that the strongest influences on policy formation are institutional or structural factors such as 'the formal rules, compliance procedures and standard operating procedures that structure the relationship between individuals in various units of the polity and economy'. Historical institutionalism thus defines institutions in a very broad sense to include formal and informal rules and procedures, including collections of norms and routines, which define appropriate actions (March and Olsen 1984, 1989). Historical institutionalism provides the conceptual framework for examining the role that institutions have played in the liberalisation of European telecommunications.

Historical institutionalists posit that the EU institutions are capable of more than mere institutional mediation and that they may develop their own impetus to change policy (Bulmer 1998, pp. 369–70). In this way, the approach sees the EU institutions as key players in the policy process and not mere arbiters of competing interests. In addition to the ECJ and the European Commission, other institutions include the European Parliament (EP); inter-institutional relations, which are often procedurally defined (for example the co-decision procedure); institutional organisation (for example the Commission Directorates-General); and institutional norms (for example DG Competition's commitment to increasing competition). These institutions, among others, have had a significant influence over policy output in the European telecommunications sector.

The supranational EU institutions, particularly the ECJ, the Commission and the EP, have indeed actively sought to contribute to the development of a comprehensive European telecommunications policy. For example, the ECJ's role in promoting integration and the constitutionalisation of the Treaty of Rome is evident in the telecommunications sector, as it is in other sectors (Weiler 1982, 1997; Burley and Mattli 1993; and Wincott 1995). In short, the ECJ consistently ruled that competition policy is fully applicable to the telecommunications sector. These were landmark rulings, which served as the platform for the Commission's subsequent liberalisation programme. In this regard, historical institutionalism helps to establish the link between the policy-making process and ECJ jurisprudence (Bulmer 1998, p. 373).

Perhaps the most distinctive concept put forth by historical institutionalists is that of path dependency, where policy choices made during the formation of an institution or the initiation of a policy continue to influence, and largely determine, the policy well into the future (Pierson, 1996, 2000; Peters, 1999). Historical institutionalists maintain that fairly stable policy patterns exist because the key policy determinants, institutions, are highly stable. Initial policy decisions become self-reinforcing; reversing them at a later date becomes unattractive due to the organisational and individual adaptations that have taken place. Dense networks of social and economic activity emerge from initial decisions, accelerating movement down a particular policy path and eliminating once-possible alternatives (Pierson 1993, 2000).

Within the telecommunications sector, at the national level, path dependency helps to explain how the Postal, Telegraph and Telecommunications Administrations (PTTs) in many Member States outlived their usefulness. Indeed, in each Member State, a monopolistic PTT had been established in the early part of the twentieth century as the sole provider of national telecommunications infrastructure, services and equipment. At the end of the twentieth century, however, these monopolies proved to be difficult to eliminate or modify significantly, at least partly because dense networks of actors had become reliant on them. As a result, the PTTs in France and Germany, among others, experienced 'lock in' and 'stickiness' (Krasner 1989).

Similarly, at the European level, path dependency helps to account for the role that the intergovernmental *Conférence Européenne des Administrations des Postes et des Télécommunications* (CEPT) continued to play into the late 1990s. Originally established by the Member States in the late 1950s as a 'flexible and light' (CEPT 1996) intergovernmental policy community independent of the EC, the CEPT's institutional momentum enabled it to outlive its usefulness and ultimately hinder the Commission's effort to harmonise national licensing procedures. As historical institutionalism explains, without effective mechanisms to eliminate inferior institutions, history is often inefficient (March and Olsen 1989; Hall and Taylor 1996; Pierson 1996).

There is, of course, also a need to account for the dramatic change that has taken place in the European telecommunications sector. Although historical institutionalists assert that institutions are often difficult to reform, they also recognise that modification may occur, though not all agree on the dynamics of change.[2] The two manners of institutional change posited most frequently by historical institutionalists are: i) large-scale change resulting from significant exogenous pressures; or ii) incremental change resulting from social conflict or political posturing (Krasner 1984; Hall 1986; Thelen and Steinmo 1992). In the European telecommunications sector, institutional change has often taken place in one of these two manners. For example, mounting technological and economic pressures during the 1970s and 1980s helped to stimulate the radical restructuring of the European telecommunications sector during the 1980s and 1990s. Additionally, the persistent efforts of Commission officials created a source of institutional dynamism and helped to transform the EU gradually into the central forum for coordinating European telecommunications policies.

Nevertheless, although historical institutionalism is helpful for examining institutional and policy change in general, it nevertheless has trouble 'explaining breaks from a path dependent line' (Gains *et al.* 2005, p. 28), as it is unable to account for the specific timing and scope of reform efforts. For example, the approach gives little guidance for assessing *a priori* whether a given event will be significant enough to prompt institutional change. Gorges (2001, p. 165) goes as far as to argue that the failure of the approach to account for institutional change is a 'major, perhaps fatal, weakness ... as it is at least as important to explain institutional change as institutional stasis'. Rather than pointing to a deficiency in the model, however, this criticism points more to the need to utilise it in conjunction with other approaches. In order to analyse more precisely the timing and scope of institutional change in the European telecommunications sector, this book will turn to approaches, discussed below in section 1.2, that incorporate a more independent roles for policy ideas.

New institutionalism combined with rational choice theory, or rational choice institutionalism, tends to define institutions more narrowly than historical institutionalism. Under this approach, institutions are the formal rules of political or social games that direct, shape, and restrict the rational actions of individual or political actors (Tsebelis 1990).[3] Applying this approach to the EU, Fritz Scharpf (1997) divides the policy-making process into two modes of interaction: positive integration and negative integration. This book evaluates the EU policy-making process and the Commission's authority to set the agenda under each mode of integration.

For measures of positive integration (examined in Part II of this book), the EU relies on a joint-decision system, where joint action relies upon nearly unanimous Member State agreement. That is, where there is no consensus over policy objectives, Member States have made policy making more difficult. For positive integration, such as measures under the treaty rules on the harmonisation of internal market legislation, the Commission has the exclusive authority to define legislation, which is to be accepted or rejected by a qualified majority vote. The Commission can thus choose any point within an area defined by the intersections of the Member States' indifference curves (Scharpf 1997, pp. 160–161). In other words, because the default rule is the *status quo*, any solution within the set of outcomes preferred by any qualified majority of Member States could be adopted. Developments in the telecommunications sector, however, demonstrate that the Commission's formal agenda-setting authority under the EC Treaty's harmonisation rules is not 'quasi-dictatorial' (Scharpf 1997), as Scharpf contends, and is significantly less than its authority under the treaty's competition rules. In particular, Part II of this book finds that the EP has some agenda-setting authority in this area, while the Member States have the power to block Commission proposals that they, collectively, do not agree with (Pollack 1997; Tsebelis 1994).

For measures of negative integration (examined in Part III), the ECJ and the Commission maintain the authority to compel the removal of national regulations that obstruct free movement within the Community. This authority to extend liberalisation with little political attention is a result of the Member States' explicit Treaty commitments to reduce trade barriers (Scharpf 1999). For negative integration, under the treaty rules on competition, the Commission has substantial formal authority to set the agenda, to select quite varied outcomes and to make a choice among various majorities (Scharpf 1997, p. 145). However, Part III of this book finds that the Commission is not 'politically unconstrained', as Scharpf (1999, p. 193) contends. Following public hostility to further integration in the early 1990s and the Community's increased concern for subsidiarity post-Maastricht, the Commission has been somewhat constrained in using its competition authority. In particular, the Commission has sought to take more account of Member State concerns before issuing Article 86 of the EC Treaty (formerly Article 90) directives. Therefore, it is often necessary to examine institutional norms, in addition to formal rules and procedures, in order to evaluate policy-making authority more accurately. The broader definition of institutions posited by historical institutionalists can be more helpful in explaining policy development than the narrow definition asserted by rational choice institutionalists.

Finally, the Commission can be an impotent agenda setter in seeking to achieve a treaty objective, where the authority to achieve that objective is not provided elsewhere in the treaty (Article 308 of the EC Treaty (formerly Article 235)), or where the only way to achieve a particular objective is to amend the Treaty itself. In both of these instances, the Commission is required to gain unanimous Member State approval of its proposals. This limitation of the Commission's power is best expressed in Scharpf's (1988) joint-decision trap, under which the requirement for the unanimous approval of the Member States restricts the development of European institutions. As participants in key decisions that require unanimity, the

Member States have greater control than other actors in policy development and in creating and modifying institutions. The joint-decision trap, Scharpf maintains, can lead to decisions in the EU that are systematically sub-optimal, particularly where the self-interested 'bargaining style' of decision making prevails. In the EU telecommunications sector, the joint-decision trap helps to explain how the Member States, collectively, have been able to prevent the Commission from establishing its desired policy line of a single licensing committee and a European Regulatory Authority (ERA).

Although rational choice institutionalism is useful in evaluating the authority of a limited set of actors within institutional constraints, the approach has difficulty in accommodating the pluralistic nature of the EU policy process; there are a large number of state and non-state actors involved, with fluid and complex interactions (Mazey and Richardson 1993). This makes the precise and parsimonious determination of the constellations of actors, which is necessary in order to apply rational choice theory, exceptionally difficult. Scharpf (1991) himself notes the problem of 'exploding complexity' which occurs in the application of his approach to numerous actors. His recommendation to simplify the representation of actors, in terms of understanding the development of the EU's telecommunications policy, would result in a loss of explanatory power.[4]

Because the effectiveness of analysis depends partly on the level of detail to which explanations are sought, historical institutionalism, as discussed above, and the actor-based approaches and Kingdon's framework discussed below, are generally found to be more useful in examining the totality of the European telecommunications sector. These approaches are more flexible and, in the author's view, more capable of accommodating the large number of actors involved and the complexity of the EU policy-making process, especially in its early stages. Rational choice institutionalism is nonetheless well suited for analysing the latter stages of EU policy making, or the systemic level of analysis, where a more circumscribed set of actors is involved.

Rational choice institutionalism does not have a particularly well-developed explanation for institutional change. As discussed above, this approach seeks to evaluate the impact of institutions on the behaviour of political actors and the development of policy. Since the analytical purpose of this approach is to examine outcomes, change is exogenous and often ignored, except as a new modelling problem once change has taken place (Tsebelis 1994; Steunenberg 1994; and Crombez 2001). This chapter now turns to other approaches that help explain the continuous process of learning and institutional adjustment and help account for the scope and timing of large-scale institutional change that has taken place in the European telecommunications sector.

1.2 EXPLAINING CHANGE IN EUROPEAN TELECOMMUNICATIONS

Although new institutionalism is a useful starting point for understanding how and why policy developed in this sector, other theoretical approaches are also needed to explain adequately the development of the EU's telecommunications policy. Public policy concepts and frameworks that highlight the dynamic interaction between ideas, interests and institutions are frequently more capable of explaining policy and institutional change than the institutionalist approaches discussed above.

Cohen, March and Olsen first put forth their concept of change in 1972, in their now famous 'garbage can' approach to policy making. According to this approach, institutional reorganisations are rarely planned and are 'highly contextual combinations of people, choice opportunities, problems, and solutions' (March and Olsen 1989, p. 80). Institutional modification, therefore, results from the reshaping and adaptation of preferences and possibilities within institutions. Based partially on the garbage can model, John Kingdon (1995) puts forth an approach to policy making that identifies several processes that must come together in order to produce policy change. Kingdon identifies streams of problems, policies and politics which flow through the system and have lives of their own. Although these streams are largely independent of one another, they must come together for a new policy to find its place on the agenda. The joining of the streams, or the occurrence of a policy window, is dependent upon: the recognition and definition of a problem, the production of a policy proposal, and the occurrence of a favourable political event. Although often difficult to predict, policy windows generally result from the emergence of a new problem, significant political change, or through the actions of a skilful policy entrepreneur.

John Kingdon's approach provides valuable insight into the timing and scope of institutional and policy change and helps account for the conditions that prompted the transformation of the European telecommunications sector. In particular, the Commission's ability to act as a policy entrepreneur, in putting forth innovative policy proposals (a new policy frame) and affecting institutional change, was frequently dependent on the appearance of a policy window. For example, it was not until Etienne Davignon became European Commissioner for Industry in 1977, that the problems facing the telecommunications sector were identified and framed in a winning way and support was marshalled for a coordinated European response. This policy window helped the Community to gain authority over European telecommunications policies. Similarly, the Commission's Competition Directorate-General took advantage of a policy window which opened in response to Europe's drive to compete in the information society, to push for the full liberalisation of telecommunications services and infrastructure in the mid-1990s.

1.3   POLICY NETWORKS: THE FORMULATION OF POLICY

An analysis focused solely on the role of institutions and institutional change, however, is incomplete; there is also a need to account for the interactions between institutions and numerous state and non-state actors involved in the EU policy-making process. Thus, actor-based models, such as the policy networks, epistemic community and advocacy coalition approaches, are drawn upon, especially in examining the early stages of the policy process, when a wide range of actors, experts and interests sought to influence proposals.

Policy-making approaches centred on the role of actors, or actor-based approaches, must be employed, in conjunction with the theoretical approaches discussed above, in order to evaluate developments in the telecommunications sector more effectively:

> What is gained by this fusion of paradigms is a better 'goodness of fit' between theoretical perspectives and the observed reality of political interaction that is driven by the interactive strategies of purposive actors operating within institutional settings that, at the same time, enable and constrain these strategies. (Scharpf 1997, p. 36)

In particular, the large-scale transformation of institutions and policies within the European telecommunications sector implies a significant role for governmental and non-governmental actors (see Mazey and Richardson 1993). Partly because one of the only formal EU treaty-based rules governing the initiation of legislation is that the European Commission has a monopoly over proposals, the structure of the policy-formation stage varies significantly from sector to sector, tends to be less rigidly defined and is, quite often, highly pluralistic.

Mazey and Richardson (1994) note that the 'rudimentary elements of a European policy style' may be developing in which an 'adolescent' and 'promiscuous' Commission oversees an agenda-setting phase characterised by 'extreme openness'. More recently, Mazey and Richardson (2001) find that the policy formation stage is becoming increasingly institutionalised as the Commission develops structures to facilitate interest consultation. Indeed, as the Commission gained a central role in telecommunications policy formation, the consultation of affected interests became increasingly institutionalised. Commission officials found that it is necessary to establish structures, whether *ad hoc* or permanent, where affected interests can articulate problems and discuss possible solutions, and where Commission officials can marshal support for the policy line that emerges. The policy networks, advocacy coalitions and epistemic communities approaches discussed below, in conjunction with insights from new institutionalism, are helpful in describing and explaining this process.

The policy networks approach is a method of research rather than a coherent policy-making theory. Policy networks are where decision makers and interests come together to negotiate solutions to policy dilemmas. Analytically speaking, it is their level of integration that distinguishes them:

> At the one end are tightly integrated *policy communities* in which membership is constant and often hierarchical, external pressures have minimal impact, and actors are highly dependent on each other for resources. At the other are loosely integrated *issue networks*, in which membership is fluid and non hierarchical, the network is easily permeated by external influences, and actors are highly self-reliant. (Peterson 1995a, p. 77)

Policy communities are thus highly integrated, stable and resource-dependent policy networks, while issue networks are instable, permeable to non-sectoral influences and have weak resource dependencies (Peterson 1995a).

Researchers have recognized the applicability of the policy networks approach to the EU. Mazey and Richardson (1993, p. 253), for example, note that the EU shows 'quite significant variations in the nature of policy networks (and that in some specific policy areas no networks may exist) but that there is at least a case to be made that the network concept is quite useful'. The flexibility of the approach enables researchers to consider the complex, multi-level, pluralistic, and frequently informal nature of EU policy making (see Peterson 1995a; Mazey and Richardson 1993; and Rhodes, Bache and George 1996). As Schneider *et al* (1994, p. 112) comment:

> The highly pluralist pattern exhibited by the EU policy networks is a consequence not only of numerous actors' efforts to influence the European policy process in an early stage of formulation, but also of a deliberate networking strategy employed by the European institutions, especially in the Commission.

Several authors consider the study of policy networks to be within the framework of new institutionalism. Peters (1999), for instance, contends that the informal and less structured clusters of actors, experts, and officials display some characteristics of institutions.[5] Rhodes (1997, p. 12) also consider the policy networks approach to be an application of new institutionalism in its broad sense.

The policy networks approach is, therefore, at the very least, highly compatible with new institutionalism. However, in the author's view, including the approach within the new institutionalist framework seems to define institutions overly broadly and results in the reduced explanatory power of both new institutionalism and the policy networks approach. Thus, by applying the two approaches separately, for example, chapter two finds that the French and German governments opened their telecommunications policy networks progressively to industry and interest groups in an effort to reduce criticism and facilitate comprehensive institutional reform. In each Member State, the sector had been dominated, since the early part of the twentieth century, by a restricted number of actors forming a closely tied policy community. During the 1980s and 1990s, however, technological and economic pressures and EU developments forced the Member States to reform their monopolistic markets and open their policy networks.

The policy networks approach has been criticised in its application to the EU due to the difficulties faced in outlining the boundaries of a commonly accepted policy network. Researchers, in other words, may face problems in determining which national interest groups and other international organisations are included in the network. Bulmer (1993) notes that the approach has difficulty incorporating the ECJ and compliance arrangements, which may encompass a separate network in each Member State. This problem, termed the 'boundary problem' by Hussein Kassim (1994), is especially true when Member States can choose between setting policy in the EU or another international organisation. This may be because one organisation is more comprehensive, specialised, or has a distinct organisational logic (supranational or intergovernmental). In this case, it is ambiguous whether both the organisations and Member States involved should be seen as equal members of the policy network.

A further criticism put forth by Kassim concerns the inability of the policy networks model to accommodate the complexity of the EU's institutional structure. Rather than pointing to a deficiency in the model, however, this criticism points more to the need to utilise it in conjunction with other approaches and, in particular, new institutionalism. Although the policy networks approach helps us to understand the policy-making process, it is only one part of the process (Rhodes *et al.* 1996). EU researchers have increasingly recognised the need to utilise a combination of approaches to analyse and explain EU policy making. For example, as Wayne Sandholtz (1996, p. 427) notes, 'different kinds of theories are appropriate for different pieces of the EU puzzle'.

This policy networks approach is the appropriate tool for examining the early stage of policy formulation. At both the national and EU levels, governmental officials seek to involve a wide range of actors to gain advice and support for the initial framing of policy proposals. This early stage of the EU policy-making process (examined in chapter four) is not governed by detailed EC Treaty rules and therefore policy networks analysis may prove more beneficial than institutional analysis in explaining this stage of the policy process.

However, the latter stages of the EU policy-making process (examined in chapter five), are governed by a complex set of EC treaty rules and therefore institutional analysis may prove more relevant than the policy networks model at this point. Similarly, in *Decision-making in the European Union: towards a framework for analysis*, John Peterson (1995a; see also Peterson and Bomberg 1999; and Peterson 2001) divides the EU policy-making process into three levels of analysis: the super-systemic or macro level, the systemic level, and the sub-systemic or meso-level. The characteristics of each stage and the appropriate approach, according to Peterson's model, are described below:

Super-Systemic or Macro Level

The super-systemic level concerns history-making decisions that 'alter the way that the EU works as a system of government'. Although this level includes the ECJ applying legalistic rationality in defining the Union's treaty powers, history-making decisions generally 'reflect distinctly political rationality: the desire of national governments to remain in power'. Peterson asserts that bargaining is primarily intergovernmental at the super-systemic level and that integration theories such as intergovernmentalism (Moravcsik 1994) and neofunctionalism (Haas 1968) are most applicable.

Systemic Level

At the systemic level, policy-setting decisions are generally taken by the Council of Ministers and COREPER. Once again, political rationality (that is, the desire to remain in power) is most often the basis for answering the question 'what should be done?' Peterson asserts that bargaining is primarily inter-institutional at this level and that new institutionalism is the 'best' theory for explaining policy development.

Sub-Systemic or Meso-Level

Finally, the meso-level is 'where many policy-shaping decisions are taken, particularly as policy options are formulated'. Key actors include

Council working groups, the Commission's Directorates-General (DGs), and committees of national civil servants and private actors. Meso-level decisions are often based on technocratic rationality (specialised or technical knowledge) and address the question: 'how do we do it?' Peterson asserts that the policy networks model is the 'best theory' for explaining policy development at this level.

Peterson's framework is a valuable tool for understanding the full dimension of EU policy making. In particular, this book confirms that there is a need to apply different theoretical approaches to different levels of the EU policy-making process. Nevertheless, it is also recognised that difficulties may arise in the application of Peterson's framework. The basis for his framework is in identifying patterns of decision making at different levels of analysis; in describing the 'normal' characteristics of each level. This forces him to use imprecise language to avoid being inaccurate. For example, in his only clear attempt to define the meso-level, he notes that it is 'where **many** *policy-shaping* decisions are taken, **particularly** as policy options are formulated' (1995b, p. 74, bold added, italics in original). The use of the word **many** implies that there are policy-shaping decisions taken outside the meso-level and leaves open the possibility that other types of decisions are taken within. The closest he gets to defining policy-shaping is that '**most** policy-shaping decisions are 'second-order' decisions which address the question: how do we do it?' (1995b, p. 74, bold added). Therefore, some are *necessarily* not. These soft definitions lead to ambiguities in the application of the framework. For example, he includes the EP's committees in the meso-level while admitting that 'they have little capacity to probe decisions made at the meso-level …'.[6] Peterson himself notes that 'the difficulties of deploying the policy networks model – or any other model – to the EU are considerable given the openness, fragmentation and general "messiness" of its meso-level' (1995b, p. 87).

Peterson's framework is based on two factors: the characteristics and aims of the actors involved in the policy-making process and the organisational and structural aspects of the policy-making process. In the author's view, incorporating aspects of the former makes distinguishing levels of analysis difficult. This is precisely because of the fluid nature of EU policy making. On the contrary, in a framework based entirely on organisational and structural features, identifying different levels of analysis is relatively easy. This is because structural and organisational features are generally more stable than the characteristics and aims of actors involved in policy making. The assumptions of new institutionalism, therefore, may form a more suitable basis for the construction of a theoretical framework for EU policy making.

Other actor-based approaches, which are also centred on explaining the early stages of policy making, include advocacy coalitions and epistemic communities. In focusing on clusters of actors, these approaches share similarities with the policy networks approach. An advocacy coalition is a network of actors, united by a core set of beliefs, which lobbies for, or against, a particular policy position. Sabatier (1987, p. 652) explains that advocacy coalitions contain 'actors at various levels of government ... as well as journalists, researchers, and policy analysts ... who share a set of normative and causal beliefs and who often act in concert'. This approach finds that another group, termed policy brokers, usually mediates the incompatible strategies from different advocacy coalitions. The primary objective of policy brokers is to develop a compromise position that is reasonable and will reduce conflict. In the telecommunications sector, competing advocacy coalitions centred around users, on the one hand, and the former monopolistic telecommunications operators, on the other hand. Both groups sought to influence EU policy, while the Commission acted as a policy broker between the two groups.

According to Eberlein and Grande (2005, p. 101), the 'most important resource for informal co-ordination through networks is neither law nor money, but *information*'. An epistemic community is a group of experts that participate in policy making by supplying policy makers with detailed knowledge over particular issues (Haas 1992; Zito 2001). Haas (1992, p. 3) explains that epistemic communities contain 'networks of professionals with recognized expertise and competence in a particular domain and authoritative claim to policy relevant knowledge within that domain or issue area'. Epistemic communities play a powerful role in the telecommunications policy-making process. The highly technical nature of the sector has forced Commission officials to rely on epistemic communities for information, analysis and advice. As a consequence, Commission officials have also increasingly delegated responsibility to them.

1.4   APPROACH TAKEN IN THIS BOOK

This book agrees with recent analyses, which have concluded that there is a need to incorporate more than one theoretical approach in order to explain the complex EU policy-making process adequately. It finds that insights from new institutionalism serve as a useful starting point for analysis; institutions create and constrain policy options and influence the ability of key actors to control or determine policy outcomes. Sectoral analysis, therefore, should begin with an examination of the prevailing rules and procedures. This book also finds that insights from Kingdon's framework and actor-based explanations, such as the policy networks approach, must be employed to gain a fuller understanding of the dynamics of the policy-making process. This leads the following questions to be asked in each chapter:

What is the prevailing institutional arrangement? Who are the key actors during the period examined? To what extent does this period exhibit institutional and policy stability or change? What factors best explain this stability or change? How do the institutional arrangements affect the ability of key actors to influence policy outcomes? To what extent are the policy-making approaches examined in chapter one capable of explaining outcomes?

Chapters two through nine seek to answer these questions through seven 'snapshots' of the telecommunications policy-making process in Europe from 1957 to 2003. Breaking down the policy-making process into a series of smaller case studies enables the analysis of policy making in a variety of institutional settings.

The central hypothesis of this book is that the application of new institutionalism is a useful beginning point for analysing the EU policy-making process, but that, on its own, it does not adequately account for how policy developed in the sector. This book explores various theoretical avenues for overcoming new institutionalism's weaknesses, including its inability to account for the timing and scope of the widespread institutional and policy change that has taken place. Indeed, there are many factors affecting the development of telecommunications policies and there are many more incentives for breaking from a path dependent trajectory than allowed by historical institutionalism (Gains *et al.* 2005). Although reform in European telecommunications has been spurred on continually by exogenous pressure from technological and economic developments, telecommunications policies have also been subject to many other influences, pressures and policy-making procedures.

The magnitude of change in European telecommunications makes it impossible in this book to review all developments in the sector; instead, discussion is limited to a handful of the most important regulatory developments. The book begins its examination of European telecommunications in chapter two with a discussion of policy making at the national level, which historically had been dominated by closely-knit policy communities. At the centre of each national policy community was a Postal, Telegraph and Telecommunications Administration (PTT), which maintained a monopoly over national telecommunications networks and services. Chapter three tells the early storyline of EU policy development in telecommunications, which begins with the Commission's informal agenda setting in the late 1970s when it began to take numerous small steps to institutionalise a European telecommunications network and develop an autonomous and active telecommunications policy. It funded external studies to give credibility to its policy line; warned of the threat of US domination; and used other actors, such as industry and users, to help it gain authority, by drawing those groups into the process.

By the mid-1980s, the Commission had gradually extended its competence in the sector by placing itself at the centre of an increasingly complex policy network. With the publication of a discussion document, the 1987 Telecommunications Green Paper, the Commission put forth an ambitious programme to harmonise (Part II) and liberalise (Part III) the sector. The harmonisation of European telecommunications legislation is the subject of Part II of the book (chapters four through six), where the shift of telecommunications regulation from the national level to the European level is discussed. The Commission justified the harmonisation of access conditions to national telecommunications networks and services at the European level by the perceived need to establish a level playing field for new market entrants and to facilitate the provision of new, innovative services. The liberalisation of European telecommunications markets is the subject of Part III of the book (chapters seven and eight). Part III discusses the Commission's effort to coordinate the elimination of national PTT monopolies at the European level and to introduce competition throughout the sector. The Commission justified liberalisation by the perceived need to increase the quality and lower the cost of telecommunications services and increase the competitiveness of European industry. The Commission's effort to coordinate the EU's approach to telecommunications culminated in the adoption of a series of 22 directives that aimed to harmonise and liberalise the sector by 1998 ('the 1998 package').

As the following chapters demonstrate, different policy-making procedures were used to develop these directives and each directive was subject to different forces and pressures. Notwithstanding various setbacks and difficulties, in little more than a decade after the 1987 Green Paper was published, the Commission was able to help establish a comprehensive regulatory framework over telecommunications. As a result, competition has been introduced into nearly every aspect of the sector and hundreds of millions of European telecommunications users now have access to higher quality and less expensive telecommunications services. In coordinating the liberalisation and harmonisation of European telecommunications networks and services, the Commission has accomplished what would have been inconceivable to many people in the 1980s. Almost universally, the 1998 package of telecommunications legislation has been heralded as a great success.

This book tells the story of the liberalisation and harmonisation of European telecommunications policies. Its focus is on the development of some of the most important directives of the 1998 package, which helped to promote the dramatic transformation of European telecommunications. Chapter nine also discusses some of the important developments in the sector that have taken place subsequent to the 1998 package. In particular, technological developments, particularly the convergence of the broadcasting, telecommunications and IT sectors, prompted the need for further changes to the regulatory framework. As is demonstrated throughout the book, technological and economic developments have been a consistent exogenous pressure for change in European telecommunications and, to some extent, have driven policy development. Recognizing this, in 2003, European policy makers put into place an electronic communications framework that is technology-neutral and applies to all transmissions networks in the same way, so that it can encompass new, dynamic and unpredictable markets.

The book is divided into four Parts: Part I provides the historical development of European telecommunications and background information for the detailed empirical work in later chapters. Parts II and III examine telecommunications policy making in different institutional settings of the EU policy-making process. These settings include the EC's normal legislative procedures, which were used to harmonise or re-regulate European telecommunications legislation by 1998 (Part II), and the EC's competition rules, which were used to liberalise the telecommunications sector by 1998 (Part III). Part IV discusses the development of the 2003 regulatory framework and then concludes with a discussion of the wider implications of the findings of the book.

More specifically, Part I of this book introduces the important actors and institutions in the UK, France, Germany (chapter two) and at the European level (chapter three). Chapter two demonstrates that particular institutional configurations in the UK, France and Germany were indeed influential in determining reform strategies at the national level. Each country has distinctive national institutional features, which provided a good basis for testing the new institutionalist assumption that institutions influence policy outcomes: the UK, for example, has a centralised constitution with authority concentrated in the political executive, while France has an executive authority that is shared between the President and Prime Minister and Germany has a particular form of federal government which ensures that minority groups hold significant veto power over constitutional amendments. This chapter demonstrates that path dependency resulted in the national PTT monopolies in France and Germany outliving their usefulness.

Chapter three, which introduces important actors and institutions at the European level, focuses on the early years, from 1957 until 1988, of

European-level action in telecommunications. It draws upon historical institutionalism and Kingdon's framework to explain the development of the EU's telecommunications policy. This highlights the way in which the Commission acted as a policy entrepreneur in seeking to mobilise support for its strategies; it created, supported and relied upon a specially created epistemic community, formed policy networks and institutionalised the policy-making environment at the European level. Legal norms were established by ECJ decisions, which further strengthened the Commission's efforts to obtain a central policy-making role in the sector.

Part II (chapters four through six) examines the EC policy-making process for the harmonisation or re-regulation of the telecommunications sector. This analysis confirms that the prevailing institutional environment under the EC's normal legislative procedures was found to be a key factor in determining policy outcomes.

Chapter four analyses the Community's normal legislative procedures at the sub-systemic level. It undertakes a detailed examination of the drafting of two pieces of harmonisation (Open Network Provision) legislation: the ONP voice telephony directive and the ONP Framework Directive. These directives were chosen because they are two of the most important directives of the 1998 package of 22 directives adopted by the Community institutions. The ONP Framework Directive was chosen because it served as the framework for much of the legislation adopted as part of the 1998 package, while the ONP Voice Telephony Directive was chosen because the voice telephony market was an extremely important revenue source for the national monopolies – it accounted for approximately ninety per cent of their profits. As such, the ONP Framework Directive and ONP Voice Telephony Directive were key pieces of the ONP regulatory framework.

Chapter four argues that new institutionalism and actor-based approaches are useful for understanding policy formulation in the sub-systemic level of the EU policy-making process. The Commission, which maintains autonomous formal authority to draft legislative proposals, established an influential policy network (the Telecommunications Committee), institutionalised a pluralistic three-month public consultation period and served as a broker between competing advocacy coalitions. By the early 1990s, after decades of policy entrepreneurship and battling with the national PTTs, the Commission finally gained a central coordinating role in the telecommunications sector. Concurrently, the Commission established an elaborate institutional structure to facilitate EC-level interest representation.

Chapter five analyses the Community's normal legislative procedures at the systemic level of analysis. It continues the detailed examination of the ONP Framework and ONP Voice Telephony Directives through the inter-institutional bargaining stage of the policy-making process. The chapter argues that, since the systemic level is governed by a complex set of EC treaty rules, institutional constraints were important in structuring negotiations. Rational choice institutionalism is particularly well suited to analysing the struggle between a limited set of rational political actors as they adapt their strategies in response to frequent institutional and procedural changes. Successive treaty modifications, for instance, have enabled the European Parliament to become a key actor in the telecommunications sector. Chapter five also discusses the possible advantages of relying exclusively on structural features to distinguish the sub-systemic level from the systemic level of analysis.

Chapter six examines the privileged position of the Member States in the EC policy-making process. The prevailing rules and procedures, along with the normative commitment to subsidiarity, enabled the Member States to resist the Commission's effort to Europeanise the telecommunications sector in key areas. Although favoured strongly by the Commission, European Parliament and industry, the Commission failed to establish a European Regulatory Authority (ERA) or a single licensing committee and faced Member State resistance to the establishment of an EU-wide licensing regime. The Member States were able to shift the harmonisation efforts of the Licensing Directive to the intergovernmental CEPT, which significantly hindered the development of an EC-level licensing regime.

Part III (chapters seven and eight) examines the Commission's coordination of the liberalisation of the EU telecommunications sector. Generally speaking, the EC treaty competition rules give the Commission substantial authority to create a level playing field in the telecommunications sector. As highlighted in this section, Article 86 of the EC Treaty (formerly Article 90) helped to shape the policy environment and affected the strategies, preferences and goals of the key actors.

The focus of chapter seven is on the 1988 Terminal Equipment Directive, which liberalised telecommunications terminal equipment (for example telephones, modems and fax machines), and the 1990 Services Directive, which liberalised telecommunications value-added services (for example electronic mail, telereservations). Article 86 was included in the original Treaty of Rome at the request of Germany and the Benelux countries; the chapter argues that historical institutionalism accounts for its unintended consequences, which led the same Member States 30 years later to protest at the Commission's use of the previously latent article. This chapter confirms that both the Commission and ECJ are key supranational actors that maintain an institutional bias towards furthering integration. Historical institutionalism also accounts for the post-Maastricht normative shift of the EU institutions towards respecting subsidiarity, which led the

Commission to retreat from its authoritative policy style; it pursued a more accommodative policy, sought Member State consensus, and refrained from using Article 86.[7]

This focus of chapter eight is on a series of amendments to the Services Directive in the mid-1990s that expanded its scope to include satellite communications, mobile telephony, voice telephony services and telecommunications infrastructure. The chapter demonstrates that, as the policy environment became more favourable, the Commission became more assertive and began to invoke, once again, its formal powers under Article 86. In line with Kingdon's framework, the Commission was an effective policy entrepreneur; it maintained sufficient resources to facilitate the spread of policy ideas and took advantage of a policy window to achieve EU-wide infrastructure liberalisation.

Part IV (chapters nine and ten) discusses the development of a new regulatory framework for EU telecommunications and the prospects for a new theoretical approach for EU policy making. More specifically, chapter nine relies on Kingdon's approach (1995) to highlight the Commission's ability to take advantage of a 'policy window' to put forth the 2003 electronic communications framework, which is adapted to the convergence of the telecommunications, information technology and audiovisual sectors. Chapter nine also discusses the Commission's development of a transnational regulatory network to meet the need resulting from the increased harmonisation of telecommunications policies at the European level and the failure to establish an ERA to oversee implementation at the national level.

The conclusion (chapter ten) discusses the wider implications of the findings of the book and explains how it contributes to the development of a theoretical approach for EU policy making. Although new institutionalism is determined to be helpful for selucidating the important role that institutions play in policy development, the approach is unable, on its own, to account adequately for the dramatic and far-reaching changes that have taken place in the EU telecommunications sector. It is concluded that a synthetic approach is likely to be a valuable – perhaps necessary – tool for understanding and explaining the multifaceted and dynamic EU policy-making process in other policy areas.

NOTES

1. For a more detailed description of three variants of new institutionalism, see Schneider and Aspinwall (2001) and Hall and Taylor (1996). Lowndes (1996) characterises the different positions of new institutionalist authors into six 'vignettes', while Peters (1999) distinguishes seven new institutionalist approaches.

2. Krasner (1984) puts forth the concept of punctuated equilibrium, in which institutional modification occurs rarely and, in almost every case, results from a crisis situation. Thelen and Steinmo (1992) agree that institutions experience instances of dramatic change, but argue that they may also evolve slowly, and that this evolution typically results from particular conflicts or posturing by actors. They also identify several sources of institutional dynamism, under which policy change can occur within the same institutional arrangement. Significant exogenous changes to the political or economic environment may enable previously latent institutions to become salient, existing actors to shift their strategies or goals, or new actors to pursue new goals more effectively. For example, they point to the political events surrounding the SEA as having transformed the European Court of Justice into an increasingly significant European institution. Hall (1986) also asserts that institutions evolve slowly and experience far-reaching changes only at crucial moments in history. Over long periods of time, Hall contends, social conflict and societal change can alter the structure of institutions. In addition, policy change occasionally 'breaks through some of the conventional constraints' (Hall 1986, p. 273). Under certain circumstances, such as the election of a new political party with new solutions, policy modification can take place without institutional reform.
3. The basic argument put forth by rational choice institutionalists is that actors are motivated primarily to maximise their utility, and that their objectives are obtained most easily through institutional membership. The approach, therefore, assumes that institutions affect only the strategies of actors; preferences and goals are determined exogenously to the institutional context. These assumptions contrast with the concepts of historical institutionalism examined above whereby institutions have a normative influence on actors and therefore 'leave their own imprint on political outcomes' (Thelen and Steinmo 1992, p. 9).
4. Scharpf (1997, p. 81) gives the example of simplifying the Council of Ministers into 'rich' and 'poor' Member States in order to examine the negotiations over European regulations for social welfare and industrial relations. This approach would ignore the subtle, yet important, differences in Member State preferences that will be highlighted in this book. In addition, only primary actors, which are participating directly in policy choices, are considered within the actor constellations. This book, on the other hand, highlights the important role played by interest groups and independent experts.
5. See Peters (1999) chapter seven on 'Institutions of Interest Representation'.
6. With the extension and consolidation of the co-decision procedure (see chapter five), Peterson would presumably agree that this description of the meso-level (put forth in 1995) is inaccurate. Nevertheless, this illustrates the dangers of basing a theoretical framework on a description of the normal characteristics of the levels of analysis. A change in characteristics necessarily results in a change to the framework, thus making it difficult to apply it consistently over time.
7. Article 86 of the EC Treaty gives the Commission the authority to address 'appropriate directives or decisions' to the Member States in order to ensure the effectiveness of the free-trade, non-discrimination and competition rules set out in the Treaty.

2. Policy making in the United Kingdom, France and Germany

A much more complex structure of policy making has developed [at the EU level], encompassing a wide range of actors. All these actors, especially national governments, are having to adjust to the reality of the situation. In a very real sense, they have all lost some power in a common pooling of sovereignty. (Richardson 1996, p. 3)

INTRODUCTION

This chapter examines the reform of national policies and policy-making structures in the telecommunications sectors of the UK, France and Germany between 1980 and 1997. The relative importance of institutions and actors in shaping reform efforts in the three Member States is assessed. It is argued that distinctive national institutions had a significant effect on reform efforts at the national level; institutions facilitated reform efforts in the UK and impeded reform efforts in France and Germany. As later chapters demonstrate, these national reform efforts had a significant effect on the scope and timing of reform efforts at the European level.

First, this chapter briefly introduces common features of the British, French and German telecommunications sectors prior to the early 1980s. In each case, policy making was stable and dominated by a tightly knit policy community centred on the national Postal, Telegraph and Telecommunications Administrations (PTTs). Secondly, the technological and economic developments which prompted reform efforts in the three Member States, are highlighted. This analysis, more generally, introduces a theme which runs throughout the book, namely the degree to which efforts at the national level to reform the telecommunications sector have been driven by the perceived need to react to external pressures and, at the EC level, to 'catch up' with the United States and Japan. Thirdly, new institutionalism and the policy networks approach are employed to identify the factors that both aided and impeded institutional reforms in the UK, France and Germany. In each case, the institutional framework was transformed and the tightly knit policy community was opened up; new institutions were created, new actors entered the process; and the central objectives of the policy changed as all three countries moved towards liberalisation.

Reform occurred most rapidly in the United Kingdom. The British system of 'responsible' government, which centralises power in the Prime Minister and cabinet (Hall 1992, p. 107), along with the dominant ideological commitment within government and industry to liberalisation, enabled the Conservative government of Prime Minister Margaret Thatcher to push through far-reaching institutional reform over some trade union opposition in 1984. Reform in France, on the other hand, was constrained by the long-standing ideological commitment to Colbertism and Gaullism, both of which endorse state intervention and *dirigisme*, along with forceful union opposition to modification of the national PTT statute, under which all employees were guaranteed civil service status. In consequence, comprehensive institutional reform did not take place until the mid-1990s when the return of the (by then neo-liberal) Right to government opened a policy window and pushed France towards liberalisation. Meanwhile, the reform process in Germany was slowed by the requirement that the 1949 Basic Law (the constitution) be amended prior to reforming the national PTT, along with opposition from powerful unions and the Social Democratic Party (SPD). Germany's 'concertative' (Dyson 1982) policy-making style, which obliged the government to pursue extensive consultations with all interested parties, including trade unions, employers and political parties, did not result in comprehensive reform until the mid-1990s.

2.1 TRADITIONAL TELECOMMUNICATIONS POLICY COMMUNITIES

From the early part of the twentieth century until the early 1980s, telecommunications markets in Europe were divided into strict, national monopolistic regimes, the Post, Telephone and Telegraph (PTT) administrations (Foreman-Peck and Müller 1988). Since telecommunications infrastructure and services were viewed as natural monopolies, the PTTs operated the national telecommunications infrastructures and maintained special and exclusive rights over the supply of telecommunications services. The PTTs acted as regulatory bodies and generally held monopoly rights over the provision of customer premise equipment (CPE).

European governments felt that it was necessary to limit the operation of the public telephone network to a single operator in order to protect the integrity and quality of the network (Mansell 1993, p. 69). In return, the PTTs were required to maintain a universal service. This consisted of providing equivalent telecommunications services (including public telephone boxes, directories and access to emergency services), to all citizens at equivalent prices throughout a country's territory. The PTTs also maintained close ties to equipment suppliers, while little room was left available for new entrants. At the start of the 1980s, therefore, the PTTs monopolised telecommunications services, networks and equipment as well as the development of telecommunications policy (Sandholtz 1992b, p. 255).

The PTT-centred institutional framework thus limited the access of actors to the policy-making process. The closed nature of the markets resulted in a highly restrictive, tightly knit policy community with stable and continuous relationships. Membership of the community was based on the shared responsibility for the delivery of basic telecommunications services. Moreover, national governments had little incentive to modify the situation prior to the 1980s, as the national PTTs contributed heavily to the financing of national budgets. As a result, telecommunications markets and network technologies were fragmented along national lines and service provision was limited to a set of standardised services. Due to external technological and economic pressures discussed below, however, a shift in orientation took place in the UK, France and Germany during the 1980s; the national governments in these countries began – at different speeds – to take steps to liberalise and deregulate their restrictive telecommunications regimes.

2.2   TECHNOLOGICAL AND ECONOMIC DEVELOPMENTS STIMULATE REFORM

Since the 1970s, telecommunications networks and services have been revolutionised by technological developments (see Thatcher 1999a, chapter 3; Commission 1987, pp. 28–43). These developments have forced governments to respond. The rapid development of microelectronics and computers permitted a tremendous decrease in costs and increase in capabilities. Digitalisation of both transmission and switching functions increased the capacity of telecommunications transmission due to higher speeds, lower resistance, greater efficiency and sharply decreased costs. Optical fibre networks provided better quality transmission, increased transmission capacity and rapidly declining costs. In addition, new transmission techniques were developed, such as microwave and satellites, which offered greater capacity, improved quality and substantially lower costs. These advances in technology led to the development of sophisticated network and terminal equipment, the blurring of basic and advanced services and drastic cost reductions.

At the same time, these developments altered the cost structure for the national PTTs and increased pressure on them to alter their pricing structures. Because of their universal service commitment, the PTTs had for many years subsidised local services and network access (connection and rental charges) with high profits earned on long distance services. Michel Carpentier (1993), European Commission Director-General for Telecommunications,[1] estimated that, as late as 1993, ECU 16 billion were transferred from profitable long distance and international services to finance local services and other PTT obligations. However, as technological developments made distance less relevant for telecommunications, long distance tariffs became increasingly unrelated to costs.[2] The inability or unwillingness of the PTTs to adjust tariffs rapidly, along with institutional stickiness, brought about increased incentives for competition.[3] As Claudio Coretti, head of the Italian PTT, commented in 1989:

> We know that this [tariff rebalancing] will take a certain time, and we do recognise that, in some countries more than in others, the imbalance at present between the local tariffs and long-distance tariffs is such that [competition] may allow a certain amount of cream skimming that may affect the financial viability of the telecommunications administrations. (Analysys Consultants 1989)

It became increasingly clear that, in a liberalised environment, potential competitors and multinational firms could take advantage of technological developments to install advanced equipment, including alternative cellular radio and satellite networks. This put the PTTs at a disadvantage, because the costs of setting up new networks decreased in line with the reduction in transmission and networking costs. In many cases, multinational firms found that it would be less expensive to construct their own private networks than to continue to rely upon national PTT infrastructures. Also, companies increasingly requested advanced services that the PTTs were incapable of providing.

At the same time, technological developments enabled users to bypass PTT monopolies. For example, users could take advantage of the latest equipment to make calls through lower-cost and more flexible networks located in countries with more liberal regulatory regimes. The Commission (1992c, p. 21) noted that:

> Automatic 'call-me-back' equipment allows for traffic diversion from high tariff operators and a subsequent loss of revenue for them. Since traffic diversion is now technically possible and does not contravene the rules of competition it is an inevitable activity where tariff conditions favour its development. Consequently the risk of revenue loss for high tariff operators will continue.

Dial-back services and other resale activities expanded dramatically as European firms took advantage of advanced technology to circumvent the national PTT infrastructure. International calls were routed increasingly through the US (and eventually the UK), where international telephone traffic incurred substantially lower tariffs (Analysys Consultants 1991, p. 33; *The Economist*, 12 September 1992, p. 101). These technological developments made the monopolistic PTT arrangements increasingly unsustainable and guided decisions in the UK, France and Germany to restructure the telecommunications sector.

Economic developments also stimulated reform pressures. In the 1960s and 1970s, European governments acted from a Keynesian worldview. The national governments established state-controlled 'national champions' in various sectors and pursued economic growth through the management and regulation of the economy. National utilities such as telecommunications, therefore, were viewed primarily as instruments of economic and industrial policy. National governments used the PTTs as policy instruments in the pursuit of the management of wages and prices, and the reduction of employment.

European governments sought to limit technological dependence on US computers and electronics by arranging national mergers, distributing subsidies and granting monopoly rights to key sectors such as telecommunications. At the same time, they sought to modernise their telecommunications networks, as various governmental commissions and reports emphasised the importance of technological advancement. The West German Commission for the Expansion of Technical Communications Systems *(Kommission für den Ausbau des technischen Kommunikationssystems)* published a report in 1976, which persuaded the government to adopt the joint 'Programme Technical Communication' of the postal and the research ministries in 1978. Similarly, the seminal Nora and Minc report was published in France in 1978, and the *'Plan Télématique'* adopted later that year. Near the end of the 1970s, Britain introduced its videotext system, Prestel, and the 'teletext initiative' (Schneider and Werle 1990; Humphreys 1986). All these initiatives were designed to cope with technological innovation.

Meanwhile, during the mid- to late-1970s, government officials began to question the effectiveness of Keynesian economic policies, following a series of dramatic increases in the price of oil that led to widespread recession and inflation. During the 1980s, it became increasingly apparent that national strategies designed to advance national technological development were largely unsuccessful. The gap in technology between the US (joined by Japan) and Europe had increased while the 'national champions' remained isolated from competitive pressure (Peterson and Sharp 1998). Markets were fragmented with the ten Member States using nine different digital telecommunications switching systems. No single Member State market, however, was large enough to develop its own system. A 1986 report by the Union of Industries of the Economic Community (UNICE) concluded that 'if the introduction of a digital telephone exchange system is to be economically viable, the system must reach 8% of the world market', yet no Member State maintained more than 6 per cent of the market (Commission 1987). The growing perception that telecommunications strategies and markets are inefficient at the national level was an important catalyst for EC-level policy development.

There was also a third exogenous pressure. During the late 1970s and 1980s, Japanese electronic firms inundated European electronics markets, expanded their European operations and forced numerous European manufacturers to withdraw. The EC reacted by imposing tariffs, anti-dumping duties and voluntary export restraints. These measures proved ineffective and simply encouraged non-European firms to establish a presence for themselves within Europe (Peterson and Sharp 1998, p. 5).

Against this background, two further US developments were influential in forcing change in European telecommunications policy networks. In 1982, the US broke up their private phone monopoly, AT&T, thus freeing the world's largest phone company to involve itself actively in European telecommunications markets. Meanwhile, IBM settled a long-standing dispute with US antitrust authorities, thus enabling the world's largest computer company to throw itself further into European telecommunications. IBM had already been a significant player, actively challenging the European PTTs over interconnection standards. The two American companies immediately sought alliances to facilitate their entrance into the European market (Dang-Nguyen 1985). AT&T joined up with Olivetti and Philips, while IBM purchased stock in AT&T's rival in long-distance services, MCI (Schneider *et al.* 1994, p. 482). Thus, two giant American companies gained a substantial foothold in Europe, opening the European door further to US competition.

During the 1980s, the US government and US companies also began to press Member State governments directly to open up their telecommunications markets (Schiller 1982, p. 149). For example, the US firm PanAmSat, which had launched the first independent intercontinental telecommunications satellite in 1988, was influential in liberalising satellite services in the UK. PanAmSat had demanded and, with the support of the US administration, gained the right to transmit directly to British earth stations (Thatcher 1999a, pp. 238–239). US cable operators also successfully lobbied the British government for the removal of conditions from their licences, thus enabling them to establish themselves as telecommunication service providers at the local level.[4] The US government also pushed strongly for liberalisation of the German market:

> [T]rade legislation in the US Congress in Washington, requiring the President to retaliate against countries that do not open their markets to US goods, specifically mentions the Bundespost monopoly as an irritant. There seems to be little doubt that US pressure has pushed the [German] Government into acting more quickly than it might have wanted to. (*Financial Times*, 2 June 1987, p. 2)

The introduction of competition into the supply of value-added network services (VANS), CPE and satellite transmission was influenced by US retaliatory threats, as well as AT&T and IBM lobbying (see *Financial Times*, 31 July 1986, p. 6; Schmidt 1991).

The international economic environment, and perceptions of appropriate policies, began to change. European policy makers were affected as the dominant policy paradigm shifted. In contrast to the Keynesian worldview in the 1960s and 1970s, British, French and German policy makers increasingly viewed deregulation and liberalisation as the best ways to improve competitiveness; the commitment to the PTT as a public monopoly declined; and even European equipment manufacturers and service providers began seeking opportunities in foreign markets. The notion that trade reciprocity would eventually eliminate trade barriers helped to stimulate the restructuring of the European telecommunications sector.

Peterson and Sharp (1998, p. 76) note, 'if Europe's PTTs [posts, telegraph and telecommunications agencies] and their "national champion" suppliers did not change of their own volition, they were going to be forced to do so'. Similarly, Sandholtz (1992b, p. 256) concluded that:

> Purely national strategies of the traditional European kind were discredited. Technological change and deregulation abroad were undermining the PTT monopolies, and governments increasingly realized that the nationalistic fragmentation of the past would not serve their present and future needs. Thus, policy-makers were in the adaptive mode, searching for new approaches to the telecommunications sector; major studies and reforms were in progress in all the EC countries by the mid-1980s.

The British, French and German governments were faced with technological and economic developments similar to those described above. These significant exogenous changes in the environment led to the transformation of the policy communities in each country. This phenomenon is most obvious in the UK, where the PTT regime was broken up in the early 1980s.

The policy communities in France and Germany, on the other hand, were more entrenched institutionally than the policy community in the UK.

Also, the French and German policy frames were more resistant as French and German policy makers were less enthusiastic about liberalisation than the British policy makers. The French and German monopolistic regimes had become path dependent, which, in turn, slowed the reform process in both countries.

As discussed below, both institutional and political factors affected the pace and timing of national reforms. As pressures mounted and more groups sought to enter the policy-making process, it became increasingly difficult to maintain closed policy communities in the three countries. Certain institutional features discussed below, however, facilitated reform efforts in the UK, and delayed reform efforts in France and Germany.

The theoretical approaches outlined in chapter one provide a framework for analysing reform strategies in the UK, France and Germany. In each case, a long period of government intervention in the sector had generated a set of norms, rules and procedures that shaped the policy and the perceptions regarding policy objectives. The entrenched policy communities, composed of restricted numbers of powerful actors, in all cases resisted modification to the established patterns of policy making. In accordance with the arguments put forth by the historical institutionalists, the prevailing institutional configurations in the UK, France and Germany helped shape the trajectory of the reform efforts in each country.

The centralised executive power of the British Parliamentary system, compared with the more fragmented German and French systems, facilitated the Thatcher government's efforts to impose radical change. Reform efforts in Germany required the support of a coalition government, a powerful Bundesrat, and the modification of the Basic Law (the constitution). In France, executive authority is shared between the President and the Prime Minister, while the rigid PTT structure was codified in the PTT statute. Moreover, the precarious cohabitation that arose in France between 1986 and 1988 inhibited the initial reform efforts of the Gaullist-led Chirac government. In addition, powerful trade unions played a significant role in slowing the reform process in France and Germany, whereas the weaker British trade unions had little effect upon the liberalisation and privatisation process during the 1980s.

2.3   RAPID REFORM IN THE UNITED KINGDOM

The 1969 British Telecommunications Act (1969 Act) created one of the most restricted telecommunications markets in Western Europe (Solomon 1986). After a brief discussion of the key actors within the highly restricted telecommunications sector, this section will examine the institutional factors that facilitated reform efforts in the UK following the election of a Conservative government in 1979.

The two most powerful actors in the policy community prior to the 1980s were the Post Office, which was the national PTT, and the Department of Industry (DoI). The two had close relations and continuous consultation on most significant issues (Thatcher 1999a, p. 113). British equipment suppliers such as GEC and Plessey also maintained considerable political and economic clout within the policy community. They entered into bulk supply agreements with the Post Office, faced few penalties for poor performance, had little incentive to innovate and often received higher prices than the market would have allowed. Although bulk supply agreements were officially abolished in the 1969 Act, the PTT continued to maintain its paternalistic relationship with these suppliers until 1984, when BT was privatised (Cawson *et al.* 1990). Initially, there were two well-organised unions in telecoms: the Post Office Engineering Union (POEU) and the Union of Post Office Workers (UPW). During the 1970s, however, they had little direct role in policy-making (Thatcher 1999a, p. 113).

The Post Office Users' National Council (POUNC) was established by the 1969 Act to represent British users. The Post Office was obliged to consult the POUNC on its major initiatives, but was not required to act on the POUNC's recommendations. Other user groups existed, but the PTT considered them to be outside the telecommunications regime and was not obliged to consult them (Cawson *et al.* 1990, p. 89). User interests, therefore, were not well represented within the traditional policy network. Significantly, the British policy community was less successful than those in France and Germany in providing advanced services and meeting the demands of users and, therefore, was perhaps more susceptible to criticism.

The Thatcher administration entered government in 1979 and embarked on a liberalisation programme that rapidly transformed the UK into one of the world's most competitive telecommunications markets. Significantly, and in keeping with the Thatcherite policy style, the traditional policy community that had governed the telecommunications sector in the previous decades was not involved substantially in the development of the reform policy (Moon *et al.* 1986). The trade unions, in particular, were disfavoured by the Conservative government and excluded from the policy process (Marsh and Rhodes 1992, p. 258).[5]

The Thatcher administration, instead, relied on an ideologically driven epistemic community for policy advice in this sector as elsewhere (for example, education, broadcasting, health). The government commissioned reports from right-wing strategists including Professor Michael Beesley and Professor Stephen Littlechild.[6] Their recommendations, along with advice from key players within the administration such as Kenneth Baker, Keith

Joseph and Patrick Jenkin, formed the basis for a new policy community. Confirming the key role that the adoption of academic ideas has played in British institutional reform (King 2005), the advice from this epistemic community served as the foundation of the administration's new approach to restructuring the telecommunications sector. POUNC Annual Reports, documenting increased complaints against the PTT in the late 1970s and early 1980s, were used as evidence by the Thatcher government to justify its policy in claiming that 'consumers' should be liberated from the 'dead hand of state controls' (Cawson 1990, p. 92, citing Department of Trade and Industry 1983).[7]

In addition, a significant amount of policy learning and policy transfer took place, as the US provided an example from which British policy makers were able to evaluate the effects of competition and regulatory methods (Moon *et al.* 1986; see also Rose 1989).[8] With a change in government in 1979, the UK experienced an abrupt shift away from Keynesianism economic policy to a government policy based on monetarism and reduced government intervention. This comprehensive ideological shift was imposed by the Thatcher government and enabled reform in telecoms to proceed more quickly in the UK than in France or Germany.

The institutional framework of the British polity and policy process facilitated the spread of the new Conservative government's ideas. The UK maintains a centralised constitution with authority concentrated in the political executive. This gives substantial authority to the Prime Minister and his/her cabinet to initiate policy change. In addition, the intensely competitive adversarial two-party system and government of one party, compared with Germany's coalition governments, increases the possibility of radical policy shifts (Hall 1992, p. 107). Therefore, once the Conservatives entered office in 1979, the new cabinet had a sufficient concentration of power to implement distinct policy innovations and reduce government intervention.

As highlighted below, efforts of the Chirac government to shape the trajectory of reform in France were substantially less successful, at least partly because of its weaker position, while reform efforts in Germany necessitated extensive consultation with divergent interests (Schmidt 1991, p. 209). In addition, powerful business interests in favour of liberalisation were better organised in the UK than in France and Germany. The concentration of financial and banking services found in the City of London, which were heavy users of advanced telecommunications services, was not present in Germany or France (Cawson *et al.* 1990, p. 84). Business interests thus constituted an additional pressure for reform in the UK; they embraced the monetarist ideology, supported the Thatcher government's programme, and were influential in bringing about liberalisation (see Hall 1986, p. 278; Hills

1986, p. 90; Thatcher 1995, p. 244). As Cawson *et al.* (1990, p. 114) concludes, the Thatcher government overrode 'traditional producer interests in the name of greater freedom of choice for *business users*' (italics in original).

Under the 1981 British Telecommunications Act (1981 Act), British Telecom (BT) was separated from the Post Office and became a public corporation. The 1981 Act also gave the Secretary of State the power to issue licences to competing service providers. This rule change was significant because, under the 1969 Act, the national PTT had hitherto been the only organisation capable of licensing competitors. Since the PTT had an interest in maintaining its monopoly and did not perceive its best interest to be served through the introduction of competition, telecommunications operations had remained under the monopolistic control of the Post Office. However, Secretary of State Keith Joseph quickly used his authority under the 1981 Act to introduce competition in customer premises equipment (CPE) and through the licensing of a second operator, Mercury, in February 1982.

The 1984 British Telecommunications Act (1984 Act) brought about more substantial reform. BT was privatised and competition was introduced into most of the sector, including value added network services (VANS). A new regulatory institution, the Office of Telecommunications (OFTEL), was formed as an independent agency to inhibit the newly privatised BT from behaving 'predatorily'.[9] Under the 1984 Act, OFTEL was given substantial authority, including wide discretion over competition and the future nature of regulation (Thatcher 1999c, p. 95). The general duties imposed on OFTEL included securing the provision of services throughout the UK; promoting the interests of purchasers, consumers and other users in the UK; and maintaining effective competition (see OFTEL 1986).

The creation of the powerful OFTEL helped to give the UK the 'most competitive, least-regulated telecommunication system in the world by 1990–91' (Thimm 1992, especially pp. 160–180). The head of OFTEL from 1984 to 1992, Sir Bryan Carsberg, and his colleagues, were successful in liberalising the UK telecommunications sector prior to EC action (Thatcher 1996). Importantly, OFTEL was the first independent National Regulatory Authority (NRA) for telecommunications to be established in Europe and has served as a model for institutional reform in other Member States. For French policy makers, OFTEL served as a valuable example of a regulator that was institutionally independent of the government (Thatcher 1999a, p. 159). The strategies that OFTEL has employed to create competition have also served as a source of policy learning for other NRAs and the European Commission.[10]

Following the government's break up of the Mercury-BT duopoly in 1991, numerous operators and service providers have entered the market. The policy-making process has also been opened, with extensive consultations taking place between the government and all licensed operators, equipment manufacturers, user groups and trade unions that wish to participate.[11] The UK telecommunications policy network now resembles a loosely integrated issue network with a large number of actors participating in policy formation, rather than the highly integrated policy community with a limited number of actors that existed prior to the 1980s. Within this issue network, BT is just one actor among many, although the company is advantaged in the process by its resources. As a British government official commented:

> It's a matter of resources as much as anything for people, operators to respond to our consultations ... I'm not trying to say that we've worked more closely with [BT] than others, but I'm just saying that they've responded more during negotiations and consultations. They've been like a louder voice, perhaps ...[12]

Meanwhile, BT has internationalised; the company concentrated initially on extending its influence in North America, but during the 1990s also sought expansion into European markets.[13]

By the mid-1980s, the institutional environment in the UK had been transformed: OFTEL and Mercury had entered the policy network, while the relationship between the British government and the national telecommunications provider, BT, had been transformed. UK efforts also served as a model for reform, leading to policy learning and policy transfer in other EC countries. At the same time, British actors played an important role in supporting the Commission's drive to introduce competition at the EC level (discussed further in the following chapters). In this way, accounts of Commission autonomy in coordinating European telecommunications liberalisation should be tempered by reference to the fundamental role that prior liberalisation in the UK played in facilitating EC-level action. The British government pushed for liberalisation and opposed setting standards at the EC level. This was in line with the traditional government promotion of the Single Market, and related to the position of British companies that had been exposed to competition already. Lobbying by British interests and BT, in particular, was important for the development of EU policy, while multi-level governance served as a catalyst for policy transfer.

Institutional analysis also underlines the difficulties that reformers faced in breaking the dominance of the national PTTs in France and Germany. Policy reformers in these countries lacked similar institutional authority and autonomy to that possessed by the Thatcher administration. Early attempts at reform during the 1970s and early 1980s in France and Germany were either modest or failed (Schmidt 1991, p. 211; Thatcher 1999b, p. 5). Even though technological and economic developments increasingly invalidated the original justification for the PTTs – a competitive market could provide higher quality, more advanced, universal services more easily than a monopolistic PTT – institutions in France and Germany proved more difficult to reform; history was inefficient.

2.4 DELAYED REFORM IN FRANCE

France maintained a closed policy community similar in nature to the one found in the UK prior to the early 1980s. Several factors examined in this section explain why France did not implement comprehensive institutional reform on the same scale as the UK, until the 1990s. After a brief discussion of the policy community found in France prior to the late 1980s, when limited reform was introduced, the factors that impeded institutional reform in France will be examined.

The policy community was centred around the DGT (*Direction Général de Télécommunications*), which was the national PTT and formally an *administration* of the PTT Ministry. Senior DGT officials worked in close cooperation with key actors in the French government including the President, the PTT Minister, the Finance Minister and, occasionally, the Prime Minister and Industry Minister (see Thatcher 1999a). Equipment manufacturers, including Alcatel, were key actors in certain decisions (Cawson *et al.* 1990, p. 122), such as those concerning the restructuring of equipment supply (Thatcher 1999a, p. 125). The French trade unions did not have an official role in the policy-making process. The DGT, as a public administration, upheld a tradition of 'neutrality' which precluded 'private' interest intervention in the 'public' interest (Dang-Nguyen 1988, p. 134). Nevertheless, when compared to other sectors in France, the unions were represented well in telecommunications (Cawson *et al.* 1990, p. 122), and were able to delay some reform due to their ability to organise public sector strikes.

French business users, on the other hand, were less well organised and less vocal than their British counterparts. French user groups were less active than British user groups because the French telecommunications network developed later than the UK's and France lacked the reform catalyst of the 'city' as in London (Cawson *et al.* 1990, pp. 122, 127). There was also less cause for French business users to be critical of their national PTT, because the DGT had developed a modernised telecommunications network through large state expenditures during the 1970s, which had been prompted by the French state's ideological commitment to Colbertism (Cawson *et al.* 1990, p. 127). Additionally, French business users were less conscious of telecommunications costs than British business users, at least partly because

France was at least five years behind the UK in the number of professional telecom managers in its business community (Cawson *et al.* 1990, pp. 127, 146). As such, user groups had virtually no regulatory power in France (Dang-Nguyen 1988, p. 134). The lack of French user group influence in the telecommunications sector is in accordance with the widespread view that interest groups are relatively weak in general in France (see Hall 1994; Wilson 1987).

The policy community in the French telecommunications sector, therefore, was highly concentrated around the DGT and a few key government officials. The participation of other actors in the policy-making process was, for the most part, either occasional or limited to minor roles (Thatcher 1999a). Moreover, as discussed below, the French policy community was more successful than the British in satisfying demand for advanced communications services. This success helped to reduce criticism and some pressure for reform.

The deep-seated ideological commitment on the part of French politico-administrative élites to Colbertism and Gaullism, which obliged the French state to pursue policies that would enhance national grandeur, spurred French officials to devote substantial resources to modernising the telecommunications network during the 1970s (see Hall 1986; Dyson 1980). The development of high technology and industrial growth was a national priority because the sector was perceived to be a vital part of French economic and political power. Although France had one of the least advanced telephone systems in Western Europe in the late 1960s, the determined efforts of successive French governments helped create one of the most sophisticated telecommunications systems in the world by the late 1980s.[14] Therefore, *étatism* facilitated some degree of reform in France (that is, modernisation), but this reform was within the constraints of the DGT framework.

Successive French governments defined the strategies of the DGT and investment in the network was increased dramatically during the 1970s through a series of *grands projets*. These were designed to develop the national infrastructure through initiatives in videotext, teletext and broadband cable systems (Cawson *et al.* 1990, ch. 6). The DGT's status as an *administration* ensured that DGT officials worked closely with the political executive in order to fulfil the latter's goals for increased investment and expansion of the network (Thatcher 1999a, p. 139). The success achieved in modernising the telecommunications network through the *grands projets* served to reduce user criticism of the closed policy community when compared to the UK. Mark Thatcher (1999a, p. 153) explains that, 'with the success of the *grands projets* of the 1970s fresh in the minds of policy-makers and the public, discussion of the DGT's organizational position (its *statut*) and monopoly grew only gradually in the mid-1980s'. This, in turn, served to delay moves towards deregulation and liberalisation.

Government intervention in the telecommunications sector stemmed from the post-war commitment to French planning, which was socialised into civil servants through the higher education system of *grandes écoles* and, particularly, the *Ecole National d'Administration* (ENA) and *Ecole Polytechnique*. The training of nearly all French senior civil servants at the *grandes écoles* ensures that they work in broadly the same direction. As Suleiman (1978, p. 29) notes, 'no discussion of the distribution of power ... [and] the policy-making process can ignore the structure of higher education in France'. The educational system leads civil servants to maintain a particular approach to policy making and feel a certain ability and responsibility to direct industry (Hall 1986, p. 279). This institutional consideration helps to account for the pursuit of a comprehensive industrial policy in telecommunications under which various *grands projets* were pursued. As a result, more extensive plans to modernise the telecommunications network were pursued in France than in the UK during the 1970s, which accounts for the delay in some pressures for reform.

In the mid-1980s, French public and private authorities became increasingly aware that the state could no longer control the modern market economy (see Schmidt 1996a; Flynn 1995). Prime Minister Jacques Chirac entered government in 1986 inspired by Margaret Thatcher's liberalisation and privatisation wave in the UK. The Gaullist-led Chirac government charged the state with being responsible for a lack of dynamism in the French economy and pledged to loosen the state's control over the DGT. Institutional and political constraints, however, prevented radical deregulation and liberalisation during the mid-1980s. The French executive authority is shared between the President and Prime Minister, and the government was in an exceptionally precarious position during *cohabitation* between 1986 and 1988. President Mitterrand and the Socialist opposition in Parliament were opposed to the government's plans,[15] especially with regards to privatisation and deregulation. *Cohabitation* thus imposed a unique constraint on Prime Minister Jacques Chirac and inhibited the achievement of substantial institutional reform.

Another element of the institutional setting that slowed the reform process in France was the DGT's status within the French administration. This administrative statute guaranteed civil servant status for all PTT employees. Any modification of the statute was fiercely resisted by France's largest labour union, the Communist-dominated *Confédération Générale du Travail* (CGT), and the more liberal *Force Ouvrière PTT* (FO). French trade unions feared that changes to the administrative statute would lead to a loss of their civil service benefits. In other words, this network of actors had come to depend on the DGT's privileged position within the civil service, thus making change more difficult to achieve. Also, unlike in the UK in the 1980s, there was strong public support for state services in France.

In view of the forthcoming 1988 Presidential elections, and since French officials were reluctant to introduce policy changes likely to provoke social confrontation and public sector strikes, opposition from the CGT and FO prompted Prime Minister Chirac to shelve the proposed 'Legislation for Competition in Telecommunication' and to instruct his Ministers to abandon their reform programme (Thimm 1992, p. 101; Cohen 1992, p. 265). Although trade unions had no formal role within the day-to-day policy-making process, they were nevertheless able to mount a powerful defence of the working conditions of the PTT labour force (Dang-Nguyen 1988). As a result, the right-wing government failed to carry out substantial organisational reform of the DGT. Instead, the name of the national PTT was changed from the DGT to France Télécom and the *Commission National de la Communication et des Libertés* (CNCL) was created under the PTT Ministry to regulate the French broadcasting sector and limited aspects of the telecommunications sector. Most of CNCL's regulatory power, however, was over the media. Regulatory power over telecommunications thus remained largely with the PTT Minister, who continued to serve as the head of France Télécom, thereby holding the dual role of regulator and network operator (Dang-Nguyen 1988; Thatcher 1999a).

Nevertheless, the exogenous technological and economic pressures examined in section 2.2 and 2.3 continued to mount. In addition, the European Commission (1987) published its *Green Paper on the Development of the Common Market for Telecommunications Services and Equipment*, which put forth a comprehensive plan for restructuring the EC telecommunications sector (see chapter three). This development was influential in enabling Prime Minister Michel Rocard's Socialist government, upon taking office in 1988, to continue the liberalisation process begun by the former government. By invoking the EC's mandate, it was easier for the Socialist government to pursue reform. The EU thus became the scapegoat for unpopular domestic policy change (Thatcher 1996; 2004). The government could argue that France needed to respond to EC initiatives and reform efforts in the UK and Germany. Multinational corporations also, by the late 1980s, increasingly demanded better international telecoms services, flexibility and improved data transmission services (Thatcher 1999a, p. 155).

These significant exogenous pressures led to the gradual opening of the restrictive policy community. To marshal support for further reform, the Socialist government initiated a pluralistic consultation process, similar to the one undertaken in Germany (as discussed below). Under the chairmanship of Hubert Prévot, a senior Socialist civil servant, the government solicited the views of user groups, trade unions, political parties, industry associations and the general public.[16] In addition, user groups, operators and manufacturers gained a role in the decision-making process through the establishment of consultative committees (Gillick 1992, p. 728).

The widespread public consultation enabled the Socialist government to avoid social confrontation and unrest (Thatcher 1999a, p. 156), while introducing limited institutional reform of the telecommunications sector. Reform in France in the early 1990s was nevertheless still far behind what had been achieved in the UK in the early 1980s. Moreover, in contrast to British Telecom, France Télécom had not been privatised, continued to be used for government objectives, and maintained a monopoly over voice telephony services and public network infrastructure. In contrast to OFTEL, the French regulatory authority was neither functionally independent of, nor structurally separate from, the government.

It was not until a policy window appeared during the mid-1990s, that comprehensive institutional reform took place in France. The return of the Right to government in 1993 resulted in a period of cohabitation that was very different from the cohabitation experienced between 1986 and 1988. Following the 1993 election, the Right had a large parliamentary majority and President Mitterand was old, ill and less powerful. Also, the Left-Right cleavage was much less pronounced. The change in government was a development in the politics stream that allowed for changed agendas and the highlighting of the new government's conceptions of problems and proposals (Kingdon 1995), and facilitated reform.

The change in government also coincided with increased pressures for comprehensive institutional reform, as exogenous factors necessitated further transformation of the institutional framework and increased opening up of the policy community. Further reform was justified as beyond the control of the French government by reference to technological and economic developments, international competitive pressures and the need to comply with EC directives (Thatcher 1999a, p. 158). A report produced for the Balladur government (Dandelot 1993) recommended further institutional reform, and the French regulator held a widespread public consultation that found support, especially among telecommunications experts, for an independent regulatory authority (Thatcher 1999a, p. 159).

Meanwhile, France Télécom wanted to create an alliance (ATLAS) with Deutsche Telekom. The Competition Directorate-General (DG IV) of the European Commission, however, refused to approve ATLAS without agreement from the French and German governments to liberalise alternative infrastructures rapidly. France Télécom management cited EC developments as justification for organisational reform and pressed for greater autonomy to prepare for competition (see Thatcher 1996, p. 189).

This policy window, or the coupling of the problems, policies and politics streams, facilitated French efforts to catch up with prior reform in the UK, movements toward reform underway in Germany, and efforts at coordinated EC-wide reform by the European Commission.

The most difficult, or stickiest, aspect of the institutional framework to modify was France Télécom's administrative statute. The trade unions, especially the CGT and FO, continued to resist comprehensive institutional reform for fear that employees would lose their status as civil servants and associated privileges. Although trade unions traditionally are not considered powerful within French policy networks (Hall 1986; Hayward 1986), and had no official role within the telecommunications policy-making process, they continued to mount a powerful defence of the working conditions for France Télécom employees. For example, a strike organised by the unions in October 1993 managed to get the participation of 74 per cent of France Télécom employees (Thatcher 1999a, p. 161). In a country with a large public sector, there was considerable public support for these employees. Prolonged negotiations between France Télécom management and the government, and then directly between France Télécom management and the unions, finally produced an agreement in May 1996 (Thatcher 1999a, p. 162). France Télécom agreed not to release any employees other than those willing to leave early, and to continue hiring employees as civil servants until 2002.

Having overcome the final barrier to reform posed by French trade unions, comprehensive reform legislation was adopted by the government in June 1996: the statute of France Télécom was reformed; competition was introduced into network infrastructure and voice telephony services; and an independent regulator, *Autorité de Régulation des Télécommunications* (ART), was established. France Télécom was privatised partially in 1997, although the government remains the largest shareholder. One of the motivating factors behind the sale was to enable the government to raise money and reduce the budget deficit in order to meet EMU conditions.[17]

Significant exogenous changes in the environment have, therefore, resulted in the transformation of the traditional policy community and institutional framework in France. The management of France Télécom and the government were able to cite EC legislation, along with other external pressures, as justification for partial privatisation and liberalisation. Although the ideological commitment to Colbertism and Gaullism, public satisfaction with the modernised telecommunications network, the nature of the political executive, and forceful union resistance to modification of the administrative statute served to delay comprehensive reform in France, the French policy network and institutional framework have been transformed nevertheless.

2.5　DELAYED REFORM IN GERMANY

Prior to the early 1990s, the German telecommunications market was regarded as one of the least open in Europe (Schmidt 1991, p. 209). In 1985, the *Economist* referred to the national PTT, the Deutsche Bundespost (DBP), as the 'Fortress on the Rhine', explaining that the 'Bundespost has no plans, secret or not, to deregulate anything. Its near-absolute control over German telecommunications is secure' (23 November 1985, p. 20). Following a brief introduction to the key actors within the policy community, the factors that impeded German policy makers from achieving comprehensive reform until the mid-1990s will be examined.

The tightly knit telecoms policy community was centred around senior DBP officials, the PTT Minister, the Minister of Finance and the Minister of Economics. These officials had to work closely with Siemens, which was the major supplier of equipment to the DBP and a dominant actor in German industry (Cawson *et al.* 1990, p. 150). The DBP Workers Union (DPG), with membership including more than 80 per cent of DBP employees, was another key actor within the policy community. As the staunchest supporter of the monopolistic telecommunications regime, the DPG's views had to be considered by every government and PTT Minister (Cawson *et al.* 1990, p. 155). The DPG's support was centred on the notion that liberalisation would worsen working conditions and reduce employment (Cawson *et al.* 1990, p. 155). The DPG had an early and influential position in the DBP decision-making process and played a pivotal role in hampering efforts to introduce competition into German telecommunications (Thimm 1992, p. 18). German telecommunications users, on the other hand, were difficult to organise. There was only one relatively small association, Deutsche Telecom eV, which represented the interests of large users.[18] In the early 1980s, business users were 'complacent to an unexpected degree', even though they faced high tariffs and cumbersome usage restrictions (Schmidt 1991, p. 213).

Thus, the German policy community was similar in nature to those found in the UK and France prior to reform. A powerful national PTT and key government officials dominated policy making in all three countries. In addition, a favoured equipment supplier was influential, while a powerful trade union was able to delay institutional reform. Although users were not key actors in the pre-reform policy-making processes of the three countries, user groups played a critical role in British liberalisation in the early 1980s. Business users in France and Germany, on the other hand, did not exert significant pressure for reform until the mid- to late 1980s. As in France, this resulted partially from the policy community's ability to provide technologically advanced services (Schmidt 1991, pp. 211, 214).

During the 1970s and early 1980s, however, the restricted German policy community was partially opened to new actors. The DBP was able to reduce criticism and divert some pressures for reform by giving users limited input into the decision-making process (Schmidt 1991, p. 214). In addition, following the criticisms of the monopolistic DBP put forth by the computer firms IBM and Nixdorf in the late 1970s, the two companies gained a role in providing some of the DBP's equipment (Schmidt 1991, pp. 212–213). Nixdorf and IBM suspended their attacks against the DBP, at least partly resulting from their increased success at winning DBP contracts (Cawson *et al.* 1990, p. 178). As previously excluded actors entered the policy community, membership within the policy network became less restrictive. Partial opening of the policy network thus enabled German policy makers to reduce pressures for reform.

Although the policy network experienced some change during this period, institutional reform of the DBP was unsuccessful (Schneider and Werle 1991, p. 106; Schmidt 1991, p. 211). This is because the trajectory of reform in Germany was shaped by an important institutional consideration: the DBP's exclusive jurisdiction over telecommunication transmission was rooted in the Basic Constitutional Law of 1949 (the Constitution). Article 87 gave the government a constitutional monopoly over the provision of telecommunications services. The rights of the government were exercised by the Ministry of Posts and Telecommunications for public services, while the DBP was the exclusive provider of public services (Haid and Müller 1988, p. 159).

Reform of German institutions rooted in the constitution is exceptionally difficult because of the joint-decision trap, under which change requires a broad political consensus (Scharpf 1988). A two-thirds majority in favour of reform is required in both the *Bundesrat*, with representatives of the *Länder* governments, and the *Bundestag*, which is elected directly. The reform process takes on the characteristics of 'bargaining' between actors and negotiated settlements, which tends to produce 'sub-optimal' policies. Effective reform, therefore, is constrained by the participants who find that the *status quo* represents 'local optima' when compared to large-scale institutional change. At the same time, German federalism ensures that minority groups hold significant veto power over constitutional amendments.

The institutional framework, which granted the DBP a monopoly over public telecommunications services, was therefore locked in by the difficulties faced in reforming the German constitution, as discussed above. Although not as stringent as the procedures for constitutional change, simple majorities in both the *Bundesrat* and the *Bundestag* are required for the production of federal legislation. The process also takes account of particular regional interests, which generally differ from federal-level interests, thus adding another layer of lobbying, negotiation and trade-offs. The multifaceted legislative process, therefore, was an additional barrier to speedy reform (see Schmidt 1991, p. 214).

As the technological and economic pressures examined in section 2.2 began to mount in the mid-1980s, the entrenched institutional framework forced the German government to seek the broadest possible consensus for reform. This contrasts sharply with the position of the Thatcher government in the UK, as discussed above, which maintained the institutional authority to push through far-reaching institutional reform in the early-1980s. In acknowledgement of this delicate political situation, in March 1985, PTT Minister Christian Schwarz-Schilling carefully selected a politically balanced twelve-member commission to consider structural reforms to the telecommunications sector (Haid and Müller 1988, p. 176).

Although the creation of this commission signalled the government's intention to restructure the DBP (Thimm 1992, p. 56), the government also clearly recognised the political limitations of reform. Within its mandate to the commission, the government noted that the commission needed to recognize that the 'Basic Law' made telecommunications and post the responsibility of the government (Thimm 1992, p. 82). From the outset, institutional factors foreclosed the option for the commission to recommend liberalisation on the scale that had been achieved already in the UK. Thimm (1992, p. 57) contends that the two-thirds majority required to reform the constitution could not have been attained due to trade union opposition. Schwarz-Schilling thus set out to ensure that the largest consensus possible would emerge in support of 'pragmatic' change, by including representatives of employers, trade unions, political parties, academia, equipment providers and the service sector in the commission (Thimm 1992, p. 56; Haid and Müller 1988, p. 176).

Germany's 'concertative' policy style is characteristic of German economic and industrial policy sectors. Dyson (1982, pp. 35–36) notes that this style:

> emphasised not just dialogue in order to generate confidence and stable, realistic expectations and to foster a global perspective, but also expertise in order to ensure the relevance and objectivity (Sachlichkeit) of that dialogue. Continuous mutual information was to enable rational and objective decisions in terms of the public interest. In this way an effective 'global steering' of macroeconomic aggregates could be achieved with a minimum of detailed governmental intervention.

The chairman of the commission, University of Munich Professor Eberhard Witte, was highly regarded for his ability to find shared ground among opposing viewpoints (Thimm 1992, p. 56). In contrast to the very narrow consultation in the UK, the Witte Commission held a highly pluralistic consultation process similar to the consultation process in France chaired by Prévot. Numerous hearings were organised throughout Germany in order to give all interested parties the chance to express their opinions. The policy community, which had hitherto been concentrated upon the national PTT, was transformed as a result (Schneider and Werle 1991).[19] Telecommunications policy making had begun to resemble a loosely integrated issue network rather than the restricted policy community that had dominated the process for most of the twentieth century.

In an effort to learn from the experiences of other countries, the Witte Commission examined reform efforts in the US, Japan and other Member States (Dyson 1982, p. 35; Haid and Müller 1988, p. 176). The Witte Commission travelled to the US on several occasions to speak with government officials, industry officials and user groups, in order to evaluate the experience of US deregulation. Prior liberalisation in both the US and the UK provided German policy makers with a significant source for policy learning and led to a large amount of policy transfer.[20] The European Commission also had a significant influence on the German reform process (see Schneider *et al.* 1994). Herbert Ungerer, head of the telecommunications policy division of the European Commission's Directorate-General for Telecommunications, was one of the experts who reported to the Witte Commission. There were also substantial communication links between the European Commission and major German actors such as interest groups, political parties, telecommunications companies and ministries. More than 60 per cent of German policy actors rated the European Commission's influence as either strong or very strong (Schneider *et al.* 1994, p. 492).[21]

At the same time, it is apparent that the German reform efforts in turn affected the reform efforts at the EC level and in France. The Witte Commission had a significant effect on the European Commission's approach and the efforts of the French government (Thimm 1992, p. 81; and Werle 1999, p. 110). This would indicate that there was an intersection of policy networks, and that both the Commission and the Member States experienced a substantial amount of policy learning during this time of uncertainty and mounting external pressures. In other words, Member State governments were not simply responding to Commission initiatives; policy learning in Europe was multi-directional, while policy networks were complex and overlapping.

In 1987, the Witte Commission produced reform recommendations. These served as the basis for a draft reform act, which was discussed at a hearing in Germany in November 1988 and involved more than 50 organisations and individuals (Schneider and Werle 1991, p. 121). Although criticised by the DPG and SPD as going too far, and the liberal party (FDP) and industry representatives as not going far enough (Schneider and Werle 1991, pp. 122–123), Schwarz-Schilling was able to steer the reform act through Parliament in the summer of 1989 (Thimm 1992, pp. 73–74).

Postreform I separated regulatory and operational functions and introduced significant competition into services and equipment, while a new body, Deutsche Telekom, maintained the monopoly over network operation and voice telephony services. In contrast to earlier liberalisation in the UK, however, an independent regulatory authority was not established, nor was Deutsche Telekom separated from the federal administration and privatised. Deutsche Telekom continued to subsidise the postal services and contribute to the federal budget, and gained the additional burden of modernising Eastern Germany's network (Thatcher 1995, p. 250). Thus, although significant reform began in Germany by the late 1980s, institutional considerations, along with the need to obtain a political consensus for change through the 'concertative' policy-making style, prevented the government from empowering an independent NRA, seeking privatisation of the national PTT and liberalising voice telephony and network infrastructure.

More substantial institutional reform took place in Germany during the 1990s as a window of opportunity opened in response to an event in the problems stream. In particular, the government's renewed campaign for reform stemmed from increased financial pressure resulting from the unification of Germany. The government needed to privatise Deutsche Telekom partially in order to help fund unification costs, including the modernisation of the East German telecommunications network (Werle 1999, p. 112). Reform was also linked closely with developments at the EC level, including the Commission's conditional approval of the ATLAS alliance with France Télécom. Faced with the threat of increased competition and the need to internationalise, the management of Deutsche Telekom began to push hard for privatisation (see Thatcher 1996, pp. 193–194; Schmidt 1996b). These pressures and developments formed the basis of a campaign by the CDU government to convince the SPD and DPG (whose support was needed to amend the Basic Law) of the need to liberalise further the sector and to change the administrative status of Deutsche Telekom (Thatcher 1995, p. 250). Deutsche Telekom's strategy of offering shares in the newly privatised company to its employees and guaranteeing jobs (through the retention of employees' civil servant status) helped to overcome some trade union opposition.[22]

Under *Postreform II*, Deutsche Telekom became a public corporation in January 1995 and was privatised partially in November 1996. The German government, however, remains the largest shareholder. The 1996 Telecommunications Act *(Telekommunikationsgesetz)* liberalised network infrastructure and voice telephony services and established a German

National Regulatory Authority (NRA) for Telecommunications and Posts.
Despite the independence granted to the NRA, the government maintains
a degree of influence.[23] This influence is institutionalised in an Advisory
Council, which is composed of *Bundestag* and *Bundesrat* members and
has some authority to monitor the NRA and choose its directors (Werle
1999, p. 114).[24]

Nevertheless, by 1998, the structure of the telecommunications sector in
Germany was once again similar in nature to those in France and the UK.
Because of the institutional and political factors identified above, however,
reform in Germany and France took a decade longer than reform in the
UK. The financial burdens of unification, along with the developments
at the EC level examined in the following chapters, helped the German
government to overcome these significant barriers to reform.

2.6   CONCLUSION

This chapter has examined the institutional settings of the telecommunications policy-making processes in the UK, France and Germany between
1980 and 1997. In each country, the telecommunications sector had been
dominated, since the early part of the twentieth century, by a restricted
number of actors forming a closely tied policy community and by a
dominant policy frame stressing the need for state control. Technological
and economic pressures and EU developments, in particular, changed this;
the Member States were forced to reform the sector as the international
economic environment made monopolistic markets increasingly undesirable
and technologically unsustainable.

Distinct national institutions in the UK, France and Germany provided a
good basis for testing the effect of institutions on national reform strategies.
Historical institutionalism helps account for the rapid restructuring of the
telecommunications sector in the UK and the delayed reform in France and
Germany. The UK's centralised constitution clearly affected the scope for
policy change; the Conservative government of Margaret Thatcher, which
had a groundswell of public support, maintained sufficient institutional
authority to impose liberalisation and widespread institutional reform in
the early 1980s. Meanwhile, French trade unions forcefully resisted reforms
to the civil service status of France Télécom employees. In contrast to the
UK, key actors within the French policy community had come to depend
on the organisational structure (for example, national pay structure, job
security and the trade union's role in decisions affecting employees), which,
therefore, made change more difficult to achieve. As a result, institutions in

France were sticky. Similarly, in Germany, the DPG resisted modification of Article 87 of the Basic Law. The cumbersome procedures for amending the Constitution and producing Federal legislation inhibited change. In line with the insights of historical institutionalism, therefore, institutional frameworks in the UK, France and Germany significantly shaped the trajectory of reform – structural variations helped determine the distinctive paths followed in each country.

Nonetheless, historical institutionalism and the policy networks model have difficulty in accounting for the precise timing of national reform. For example, the UK, France and Germany were faced with similar technological and economic developments. Since such pressure is the only variable for explaining large-scale change in the policy networks model and historical institutionalism, these approaches have difficulty in determining when there is sufficient exogenous pressure to produce comprehensive reform and difficulty accounting for the timing of British, French and German reforms. Instead, multi-dimensional policy learning and Kingdon's framework, in focusing on policy windows and streams of politics, policies and problems, when combined with institutionalist assumptions, can help to account more precisely for the timing of comprehensive reform. In France, for example, a policy window was opened by an event in the political stream (that is, when the Right returned to power in 1993) thus enabling France to achieve the widespread reform of the telecommunications sector.

As discussed in subsequent chapters, rapid reform in the UK enabled British actors to play a key role in supporting the Commission's drive to introduce competition at the EC level. Because British companies had already been exposed to competition, the British government, along with British interests such as BT, pushed for liberalisation. Since reform in France and Germany was delayed, fewer policy actors in these countries were prepared for competition and, at least initially, supportive of EC-level action. As a result, it was not until the early to mid-1990s that both the French and German governments, as well as France Télécom and Deutsche Telekom, had shifted from passively accepting the Commission policy line to actively supporting the full liberalisation of telecommunications. The other Member States, with the exception of the Scandinavian countries and the Netherlands, proceeded even more slowly in their liberalisation efforts, although all Member States had either begun initiating changes or were reviewing the situation by the mid-1980s (see Commission 1987, p. 12). The following chapters evaluate the role of national interests, particularly those of British, French and German actors, in the increasingly institutionalised telecommunications policy-making process at the European level.

NOTES

1. During the early 1990s, Michel Carpentier was the Director-General of the DG for Telecommunications, Information, Industries and Innovation (DG XIII) (renamed DG Information Society in 1999).
2. Distance had little effect on satellite costs and was less important for optical fibre (relative to equipment) (see Thatcher 1999a, pp. 53–55; and Commission 1987, pp. 29–30).
3. Bringing tariffs closer to costs required significant institutional reform. Alternative methods needed to be found to fund universal service costs and offset contributions to national fiscal policy, as well as deal with significant overstaffing in the PTT administrations.
4. As discussed below, the Thatcher government was sympathetic to the US and, indeed, found the US to be a useful ally in its own liberalisation programme. See Gillick (1991).
5. It has been argued that one of the motives for liberalising the market and privatising BT was that Mrs Thatcher wanted to punish the monopolies and the trade unions, which she thought to be too powerful. See Marsh (1991).
6. Michael Beesley authored *Liberalization of the Use of British Telecom's Network* (London: HMSO, 1981) and Stephen Littlechild authored *Regulation of British Telecommunication Profitability* (London: Department of Industry, 1983). See Hills (1986) and Vickers and Yarrow (1986).
7. POUNC recorded that complaints about excessive delays in the provision of service had virtually doubled between 1978 and 1979, and doubled again in the following year. See Cawson (1990, p. 92), citing *Annual Reports for 1979–80* and *1980–81* (London: POUNC, 1980, 1981).
8. For a discussion of British policy learning in another sector, see Dolowitz (2000).
9. 'Predatory' behaviour occurs when a monopoly participates in a newly liberalised market and utilises profits from monopoly activities to subsidise losses resulting from aggressive pricing in liberalised sectors. See Foreman-Peck and Müller (1988), p. 34.
10. Interview, German NRA representative, Brussels, April 1999; interview, DG Information Society (DG XIII) official, Brussels, March 1999.
11. Interview, Department of Trade and Industry representative, London, May 1999; interview, DG Information Society (DG XIII) official, Brussels, March 1999.
12. Interview, Department of Trade and Industry representative, London, May 1999.
13. Interview, BT representative, London, January 1998.
14. Approximately 95 per cent of local transmission was digitised by 1990 (and 100 per cent by 1995), making France the country with the greatest availability of digital transmission. See *Financial Times*, 19 April 1990, Survey Section, p. IV.
15. Moreover, Prime Minister Chirac's Parliamentary majority was only three (excluding FN Deputies), which made his position even less secure.
16. The wide scope of the consultation process, which utilised video transmissions, Minitel and numerous hearings and debates, is detailed in *Le Deroulement du Débat Public* in Prévot, H. (1989).
17. Interview, France Télécom Representative to European Institutions, Brussels, July 1998.
18. Full membership was restricted to businesses that spent more than one million DM annually on telecommunications (Weiss 1980, p. 228).
19. In an examination of the policy network surrounding the German reform effort, Schneider and Werle (1991, p. 121) found that the number of affected or interested actors had increased significantly in comparison with previous administrative and legislative processes in telecommunications. In addition, 'the "inner circle" of influential actors became relatively large and rather "pluralist"' (p. 133).
20. Thimm (1992, p. 57), for example, notes that the experiences of the US and the UK 'had a striking impact on the commission's deliberations'.
21. During this time of uncertainty, it is likely that the European Commission influenced the policy-making processes of other Member States. See Sandholtz (1998, p. 153) (commenting on Schneider *et al.*'s 1994 study).

22.　Although 34,000 jobs were estimated to be lost, most employees could not be fired due to their civil servant status. To reduce employment, early retirement packages were offered. In addition, since becoming a public corporation in 1995, new employees no longer received civil servant status. See 'Deutsche Telekom to cut 33,900 jobs by 97', *Reuters News Service*, 29 April 1996; see also *Financial Times*, 14 December 1995, p. 28; and 'We're making a 180-degree turn', *Business Week*, 21 November 1994.

23.　The appointment of Klaus-Dieter Scheurle, a senior civil servant in the German telecoms ministry, to head the new regulatory agency was seen by many as a sign that the agency would favour Deutsche Telekom's interests. See 'Germany's new regulator is "bad choice" for entrants', *Communications Week International*, 14 July 1997.

24.　Werle (1999, p. 126) also notes that the Advisory Council adheres to the convention of parliamentary oversight of telecommunications, which was rather symbolic and not very tight.

3. The institutionalisation of EU telecommunications

The study of history does more than furnish the facts and enable us to make or test generalizations. It enlarges the horizon, improves the perspective; and ... [makes us] aware of a relationship between apparently isolated events. We appreciate ... that the roots of the present lie buried deep in the past, and ... history is past politics and politics is present history. (Sait 1938, p. 49)

INTRODUCTION

This chapter examines the institutionalisation of an EU telecommunications policy from 1957 to 1988. Analysing the construction of policy issues and associated institutions is a key objective of the historical institutionalist approach, because decisions made in the early stages of development affect later policy outcomes. Likewise, historical institutionalism contends that a great deal of policy making is 'iterative and incremental' (Armstrong and Bulmer 1998, p. 56). That is, patterns uncovered in the early years of a policy-making process are likely to share similarities with subsequent cycles of policy making.

During the period examined in this chapter, the Commission was a policy entrepreneur and, from the outset, actively sought to extend its authority over the telecommunications sector. The Commission strategically pursued its goals; it created, financed and relied upon an epistemic community, mobilised interests to support its strategies, and increasingly institutionalised the policy-making environment at the European level. The Commission's campaign to establish control over the sector involved the Commission battling with most national PTTs which wanted to minimise EU encroachment upon their work. The European PTTs had established the *Conférence Européenne des Administrations des Postes et des Télécommunications* (CEPT) in 1959, as an intergovernmental policy community with consensual, non-binding agreements over limited areas of policy. Although the CEPT monopolised the coordination of cross-border European telecommunications issues from 1959 until the early 1980s, the Commission had begun to establish a comprehensive, pluralistic, supranational regulatory regime, with binding rules over broad aspects of the sector by the mid- to late-1980s.

Kingdon's framework (1995) helps explain the Commission's ability to move telecommunications issues higher up the EC agenda and helps account for the timing of reform; policy construction at the EC level was dependent upon developments in the politics and problems streams. Although the Commission began seeking competence during the 1960s, it was not until Etienne Davignon became the Commissioner for Industry in 1977 (a development in the politics stream), and the external pressures examined in chapter two began to mount in the late 1970s and early 1980s (a development in the problems stream), that the Community began to gain legitimacy in European telecommunications policies. The first draft EC telecommunications legislation was produced in 1980 and, by the mid-1980s, the Commission had begun to establish itself as a key institutional actor in the European telecommunications sector. The Commission's landmark 1987 *Green Paper on the Development of the Common Market for Telecommunications Services and Equipment* laid the groundwork for the Commission's two-pronged strategy to coordinate the liberalisation and harmonisation of European telecommunications.

3.1 THE CEPT AND THE EARLY COORDINATION OF TELECOMMUNICATIONS

There are several reasons why the six founding Member States chose not to include the telecommunications sector within the original competencies of the EEC as established by the 1957 Treaty of Rome. First, as chapter two explained, telecommunications issues were viewed by the signatories as being primarily a national concern; policy-makers felt that the PTT was a natural monopoly which required total control over the telecommunications network in order to maintain a functional universal service. Second, telecommunications networks were, for the most part, limited to the provision of voice telephony services because technology did not enable more advanced services. As such, telecommunications were not nearly as important in economic terms as they are today; businesses were not reliant on the seamless, instantaneous, cross-border transmission of information. Third, the French government opposed further extension of supranational powers to the EEC. Proposals at that time to extend EEC authority to coordinate postal, telegraph and telecommunications issues met with strong resistance from French policy makers. Finally, the original EEC Member States also felt that it was necessary to include non-EC Member States, such as the UK and Switzerland, within a European telecommunications organisation. Britain, in particular, was considered to be an important actor in international telecommunications (Schneider and Werle 1990; Labarrère 1985).

Rational choice institutionalism is arguably more powerful than historical institutionalism in analysing the initial construction of the intergovernmental *Conférence Européenne des Administrations des Postes et des Télécommunications* (CEPT) by the national PTTs in 1959, in place of the supranational EEC, for the coordination of European telecommunications policies. For historical institutionalists:

> formative choices appear to reflect the particular confluence of political forces at play at the time of the formation of the institution. The historical institutionalist often does a good job of describing those political forces and the manner in which they produced the initial policy decisions, but that is more the product of politics than the conscious design of policy or the design of government institutions. This approach appears to eschew any rationalistic design in favor of a more political conception of policy choice. (Peters 1999, p. 72)

A central concern of rational choice institutionalism, on the other hand, is on the way in which actors seek to create an institutional design that is most favourable to their perceived interests (Peters 1999, p. 58). As discussed below, rational choice institutionalism offers valuable insight into the construction of the intergovernmental coordinating mechanism, the CEPT.

As rational actors seeking to maximise their powers, the PTTs preferred to design an intergovernmental institution, entirely independent of EEC involvement, that would enable them to coordinate a minimum amount of transnational telecommunications issues (for example, tariff principles and long-range planning) while preserving as much autonomy as possible (CEPT 1996). Thus the PTTs sought to establish an institution that was most favourable to their interests; they wanted to preserve their authority over telecommunications decisions within their respective national boundaries. In June 1959, nineteen West European PTTs thus formed the CEPT as a 'flexible and light' (CEPT 1996) intergovernmental policy community to discuss transnational telecommunications issues. All decisions were taken as recommendations, so that the competencies of the national PTTs were unimpeded (Esser and Noppe 1996, p. 552). The national PTTs were content with this institutional design because they maintained sovereignty within their territory. Up until the 1970s, when the technological and economic pressures discussed in the previous chapter began to mount, PTT ministers viewed the coordination of national activities within the EEC's rigid structure as a limitation on their freedom of action rather than as an essential alliance that would facilitate control possibilities not possible in national frameworks (Schneider and Werle 1990, pp. 86–87). In line with rational choice institutionalism, therefore, the national PTTs preferred the soft institutional design of the cooperative CEPT to the more rigid design of the supranational EEC.

CEPT membership is open to all European PTTs. Its Plenary Assembly has met every two to three years since its establishment to define the work plans of several committees and working groups (CEPT 1996; Labarrère 1985; Ungerer and Costello 1989). Initially, the CEPT was involved only in technical, operational and administrative activities (CEPT 1996). Driven, at least in part, by technological innovation, the responsibilities, size and power of the CEPT increased during the 1970s and 1980s; more extensive coordination between the PTTs was necessary to deal with the tremendous growth and development of transnational communications. In the mid-1970s, for example, the CEPT began to develop consensual interconnection standards (Schneider *et al.* 1994, p. 481). Since the early 1980s, the CEPT has become involved in the coordination of regulatory and legislative tasks. At the same time, in line with the explosive growth and importance of transnational communications, the membership of the CEPT expanded to include 26 European countries by 1989 and 46 countries by 2003.

Although the Commission has described the CEPT as 'the Community's major partner in the development of telecommunications policy', (Ungerer and Costello 1989, p. 132), during the Commission's campaign to establish control over the sector, the two institutions have frequently battled with one another over their respective competencies. The Commission and CEPT have had to work closely together

> given the [CEPT's] virtual monopoly on all practical matters for European telecommunications co-operation ... The [Commission] has not, however, had the influence that it would like to direct the CEPT into actions that would further [Commission] policy; it is an inherent weakness of the [Commission] that it has no executive action over European telecommunications services (Analysys Consultants 1989, p. 4).

The struggle for authority between the two institutions is a theme that will be examined further in this chapter and returned to in chapter six.

3.2   INITIAL COMMISSION EFFORTS TO FOSTER AN EC POLICY

Since the late 1960s, the European Commission has been determined to play an active role in the telecommunications sector. A primary rationale behind the Commission's entrepreneurship is that, like all young bureaucracies, it has an institutional interest in policy expansion. As a rational actor, it seeks to increase its own power, resources and legitimacy.[1] The Commission's first attempt to act as a policy entrepreneur was its 1968 proposal to create a committee within the Commission to coordinate the harmonisation of postal and telecommunications technical standards. The Member State governments, however, were not prepared to institutionalise policy-making authority at the EC level and preferred that European-level discussions in telecommunications remain within the intergovernmental CEPT. Following several years of unsuccessful negotiations in the Council of Ministers, the Commission withdrew the proposal in June 1973 (Schneider and Werle 1990, p. 87).

In a sense, although the Commission had developed a viable policy proposal, there was no 'problem' to be attached to its 'solution'. The exogenous economic and technological pressures examined in the previous chapter were not yet severe enough to be viewed by national policy makers as a 'problem'. Similarly, the proposal did not have the requisite political backing. As Kingdon (1995, p 202) explains, proposals that are not backed politically are less likely to move into position for a decision than proposals that are backed politically. Accordingly, the fact that the Commission's initial proposal in telecommunications was unsuccessful can be explained by the fact that the Commission did not have an open 'policy window' – the convergence of its idea (the proposal) with a problem and a favourable political event – that would enable it to push the proposal through the policy-making process.

Nevertheless, after this initial failure, the Commission did not retreat from its attempt to institutionalise an EC-level telecommunications policy. From the early 1970s, the Commission sought to foster the development of the European telecommunications sector through the opening of public procurement supply contracts. The Member State governments, however, failed to respond (Ungerer and Costello 1989, p. 129). Thus, the telecommunications sector was one of three sectors excluded from a 1976 directive that liberalised public procurement throughout the EC.[2] Significantly, telecommunications were not viewed to be within the competencies of the EC. Since the Treaty of Rome did not refer to telecommunications explicitly, the Commission had no authority to propose legislation. The EC, however, did have legal resources in trade policy. The Commission thus had to be a skilful entrepreneur and rely on its trade policy authority in its early efforts to expand into telecommunications (Schneider and Werle 1990, pp. 87–88).[3]

As discussed in the previous chapter, the telecommunications sector became more relevant for industrial policy in all Member States during the 1960s and 1970s as European governments acted from a Keynesian worldview. At the same time, consistent with theories of bureaucratic expansion, the Commission sought to expand its sphere of competence by establishing a role for itself in the area of research and technological development.[4] Even these early efforts to expand its authority, however, brought the Commission into conflict with the CEPT; the latter was concerned that the Commission sought to initiate research activities that overlapped with CEPT responsibilities (Schneider and Werle 1990, pp. 88, 90). To clarify the relationship between the two organisations, the CEPT held a conference in March 1974 that concluded with a recommendation for the EC to delegate research activities to the CEPT (Schneider and Werle 1990, p. 90). Nevertheless, as a result of the Commission's minimal authority, and at least partly due to the CEPT's resistance, the initial efforts of the Commission to extend authority in research and technological development were largely unsuccessful.[5]

Thus, despite early Commission attempts to expand its authority, institutions for the coordination of European telecommunications policies remained fairly stable between 1957 and 1977. Seeking to preserve their autonomy, the national PTTs continued to resist movement towards increased supranational coordination of the sector. The lack of desire on the part of the PTT Ministers for a substantial EC role is best illustrated by the fact that there was a 13 year gap between their first official Ministerial meeting at the EC level (to discuss postal tariffs) in 1964 and a second meeting in 1977 (see below) (Schneider and Werle 1990, p. 87; Sandholtz 1998, p. 148). Since the Treaty of Rome provided no specific legitimacy in telecommunications, the Commission needed the unanimous approval of Member States, as is required by Article 308 of the EC Treaty (formerly Article 235), for all proposals. Commission attempts in the 1960s and the early 1970s thus failed to extend significant influence into the sector.

The CEPT institutional framework for European level coordination was thus locked in and difficult for Commission officials to modify. This arrangement also privileged national PTT interests. In recommending that the EC delegate research to the CEPT, the national PTTs nevertheless recognised that the CEPT no longer had an absolute monopoly over European telecommunications issues. Because the Commission had begun to establish a small role in the sector, the CEPT acknowledged the need to try to establish limits to the Commission's role.

3.3 DAVIGNON AND COMMISSION ENTREPRENEURSHIP

In 1977, the politically savvy Etienne Davignon became the European Commissioner for Industry and Commission Vice President and became an important policy entrepreneur. Commissioner Davignon understood the challenges facing the European technology sector (Sandholtz 1992b, p. 226). Although the most immediate concern facing European nations had been the overproduction of steel, Davignon foresaw the need for a coordinated approach to raise the technical level of Europe.[6] Through his role as Commissioner from 1977 until 1984, Davignon was able to move telecommunications, electronics and computing issues into a prominent position on the EC's agenda.[7]

A window of opportunity was thus opened in the political stream when Davignon entered the Commission. Commissioners are included in Kingdon's (1995, p. 199) cluster of visible participants in the policy-making process who 'receive considerable press and public attention'. Kingdon explains,

> the chances of a subject rising on a governmental agenda are enhanced if that subject is pushed by participants in the visible cluster ... [they] do not necessarily get their way in specifying alternatives or implementing decisions, but they do affect agendas rather substantially. *Ibid.*

In his new position, Commissioner Davignon was able to bring increased attention to the issues concerning Europe's technology sector. At the same time, by the late 1970s, increasing technological and economic pressures were forcing European policy makers to respond (see previous chapter). These exogenous developments constituted a problem for all Member States, for which the Commission proposed a coordinated response. Because Commissioner Davignon could now connect proposals in telecommunications to an important problem, the possibility that they would rise on the EC's agenda was 'markedly enhanced' (Kingdon 1995, p. 198). Thus, developments in the politics and problems streams enabled the Commission to begin setting the agenda informally; Davignon claimed that the establishment of an EC-level telecommunications policy was the solution to a pressing problem.

In December 1977, the second formal meeting of the PTT Council of Ministers witnessed the beginning of a significant, independent EC-level telecommunications policy (Esser and Noppe 1996, p. 552). However, although the PTT Ministers grudgingly acknowledged the need to discuss telecommunications issues with Commissioner Davignon, they resisted his attempt to establish an extensive EC role in telecommunications (Schneider and Werle 1990, p. 90). Recognising the limits of what was politically feasible at the time, and lacking a treaty base for EC competence in this sector, Davignon suggested that the computer and microelectronics industries (which did not involve the national PTTs) receive the attention of the EC prior to telecommunications.

Davignon reasoned that progress in each of these areas would have been slowed if the Commission's initial attempt in telecommunications had failed (Sandholtz 1992b, p. 226). He also felt that, in any case, the telecommunications sector would benefit from the advances in microelectronics technologies (Sandholtz 1992b, p. 226). Davignon thus proposed that representatives from industry, the national governments and the Commission jointly work out an EC telematics strategy. With regard to telecommunications, Davignon recommended that the CEPT play the central role in developing digital networks, common standards and the harmonised provision of new services, thereby avoiding conflict with the PTTs (Esser and Noppe 1996, p. 553; Sandholtz 1992b, p. 226).

In order to marshal support and gain legitimacy for his strategy, Davignon proceeded to create and support an epistemic community of telecommunications experts at the EC level. Davignon recognised that the Commission needed a detailed knowledge of telecommunications if the EC was to gain a foothold in coordinating policy making between the Member States. Thus, Davignon institutionalised an Information Technologies Task Force (ITTF) within the Commission in 1978, principally to justify an EC role in telecommunications.[8]

Composed of approximately 70 industry professionals hired on temporary contracts (from companies such as Siemens and Olivetti), ITTF worked independently of the Commission's Directorates-General (DGs).[9] With the assistance of technical reports produced by external consultants, ITTF gained proficiency in the intricate details of the regulation and technology of telecommunications (Sandholtz 1998, p. 149). Highlighting the importance of the ITTF as an ally for Davignon's strategy, ITTF drafted a report in 1979, *La société européene face aux technologies de l'information: pour une réponse communautaire*, which was largely concerned with advancing the EC's telecommunications industry.[10] The founding of the ITTF, therefore, was the first step in gaining industry support for Commission involvement in the policy area and institutionalising a telecommunications policy network at the EC level.

Davignon presented the ITTF report to the Industry Council in November 1979, and the Council accordingly asked the Commission to submit more detailed proposals (Sandholtz 1992b, p. 227; Sandholtz 1998, p. 149). In September 1980, the Commission complied and submitted to the Council its first draft telecommunications legislation. The proposals called for the opening of the terminal equipment market and 10 per cent of the network equipment market, as well as some harmonisation.[11] Although the Commission had hoped the Council would adopt a decision before the end of 1980, the Council failed to respond until November 1984 and then only with unenforceable recommendations.[12]

Morgan and Webber (1986) blame the egoisms of the nationally oriented telecommunications industry and PTTs for hampering these early efforts at establishing an EC telecommunications policy. Schneider *et al.* (1994, p. 482) accuse, in particular, the British, French and German PTTs for the failure to harmonise telematics equipment. Their national perspectives, as examined in the previous chapter, help explain their opposition. These three Member States had either just introduced, or were introducing, their own versions of videotext and were reluctant to adapt their distinctive national versions to a harmonised EC version (Mayntz and Schneider 1988). In other words, the sunk costs of their previous decisions had made policy change more costly and more difficult to achieve. Also, the PTTs did not want to lose their power; their desire to maintain the existing institutional structure is consistent with rational actor opposition. The Commission, meanwhile, had taken several cautious, entrepreneurial steps towards establishing a supranational dialogue in telecommunications and institutionalising an epistemic community of industry experts at the EC level.

In June 1983, the Commission put forth what several authors see as a breakthrough in the development of EC telecommunications policies (Esser and Noppe 1996, p. 553; Schneider and Werle 1990, p. 92). Davignon presented a document to the Industry Council that emphasised the need for a single EC telecommunications market. Demonstrating that Davignon was a skilful policy entrepreneur, the document was sensitive to the national government interests and therefore, carefully stressed that the initiative would not affect PTT responsibility for the sector, nor national governments' ability to use PTT revenues to fund state budgets (Sandholtz 1992b, p. 228). Davignon followed up this document with a recommendation to the Council in September 1983 that put forth six action lines.[13] Specifically, an EC telecommunications policy needed to:

1. set medium- and long-range goals at the Community level;
2. define and implement a common R&D programme;
3. create the future Community infrastructure through close cooperation, especially with regard to advanced services;
4. use mutual recognition of registration standards to expand the terminal equipment market;
5. equip the EC's underdeveloped regions with modern telecommunications technologies;
6. open public procurement markets.

These action lines were to be discussed by the Commission, industry and a newly formed consultative committee, the Senior Officials Group on Telecommunications (SOGT), before the submission of more detailed proposals to the Council. The SOGT, established by Davignon towards the end of 1983, was composed of high-ranking officials from the national PTTs, economics and industry ministries. Crucially, the SOGT was an important network of actors which further institutionalised interest intermediation in EU policy making. The CEPT, however, had a powerful influence over the SOGT, because the SOGT drew most of its membership from CEPT delegations (Analysys Consultants 1989). Nevertheless, although the PTTs were not always supportive of the Commission's plans (as will be seen in the following chapters), Davignon was aware of the need to bring the national PTTs within the EC as a consultative committee in order to further institutionalise the supranational telecommunications policy community. The PTTs remained powerful actors within the policy community and therefore needed to be consulted.

In line with Kingdon's framework (1995), the Commission sought to move its pet policy proposal (the six action lines discussed above) higher up the EC agenda. The ITTF began to alert industry and the Member States that the mounting external pressures discussed in the previous chapter constituted a serious threat to European competitiveness. After the Competition Directorate-General (DG IV) initiated legal proceedings against IBM in 1984 on the grounds that it was abusing its dominant position, the ITTF warned of the threats posed by the US, particularly AT&T and IBM, and Japanese firms (Caty and Ungerer 1984, p. 49; Schneider and Werle 1990, pp. 92–94).

In order to gain a better understanding of the possible consequences of Community inaction, and wider support for its strategy, the ITTF mobilised an epistemic community of external consultants. Epistemic communities such as this one are able to 'shed light on the nature of complex interlinkages between issues and on the chain of events that might proceed either from failure to take action or from instituting a particular policy' (Haas 1992, p. 15). Indeed, Commission officials have relied increasingly on epistemic communities for information and advice and have delegated responsibilities to them and used their scientifically sound studies as a central element in the Commission's strategy (see, for example, Richardson 1996a). The ITTF's epistemic community consisted primarily of consulting firms such as The Yankee Group, Arthur D. Little, Mackintosh and McKinsey. The studies they produced repeatedly raised the threat of US and Japanese dominance in the sector and supported the Commission's claim that this threat required a joint European response.

Toward the end of 1983, Davignon further institutionalised the supranational EC policy community by incorporating the ITTF into the Internal Market Directorate-General (DG III), where the ITTF used its new institutional basis to foster awareness and support for policy activity. In a series of public hearings, the ITTF presented the findings of the epistemic community of external consultants to major industry representatives (Schneider *et al.* 1994, p. 485). The ITTF then formulated a programme based on the six action lines (see above), which called for the liberalisation of some PTT equipment purchases and a coordinated approach to research. This programme was forwarded to the Industry Council in a May 1984 Communication (Sandholtz 1992b, p. 230).

Consistent with both liberal intergovernmentalist and neofunctionalist theories of integration, the Communication warned that 'the capacity to meet [the US and Japanese] challenges, and to cope in a timely manner with the opportunities born out of telecommunications, is outside the capability of national operators on their own'.[14] Dang-Nguyen argues that the Commission's ability to dramatise the US threat is the main factor that serves to explain the beginning of an EU telecommunications policy. He notes that:

> twelve years of difficult relationships with PTTs were canceled, thanks to what one would call a rhetoric of persuasion. The trick has been to get the political support of the heads of government, instead of the PTTs. But to get it, the Commission had to develop arguments akin to previous action in the crisis sectors. If the member states could be persuaded of the likelihood of collapse, they would be induced to cooperate.[15]

The ten Ministers of Industry, and *not* the PTT Ministers, finally approved the Commission programme in December 1984 to counter foreign advances in the telecommunications sector (Schneider and Werle 1990, p. 94). The common external threat of foreign competition thus helped to spur the Member States into working together towards a solution. In focusing on the self-interested nature of rational actors within institutional constraints, rational choice institutionalism offers insight into the Member States' willingness to accept the Commission's supranational solution. A common external threat or 'common vulnerability' is more likely to cause political actors to seek solutions rather than bargain (Scharpf 1988, p. 261). This problem-solving style of decision-making enabled the Member States to overcome the joint-decision trap of the EC policy-making process, which generally makes agreement on significant institutional and policy change difficult to achieve (Scharpf 1988).

In addition, the preferences of key actors in the UK, France and Germany, as examined in the previous chapter, also help account for the Council's approval of the programme. Due to the radical liberalisation imposed by the Thatcher government early in the 1980s, BT and British equipment manufacturers GEC and Plessey had been exposed to (and benefited from) competition previously and therefore supported further competition at the EC level. In France, the DGT (national PTT), which had gained in confidence due to the successful modernisation of the French network through successive government-funded *grands projects*, also favoured some liberalisation (especially in the equipment sector) (Mayntz and Schneider 1988). Finally in Germany, after computer firms such as IBM and Nixdorf began winning contracts to supply equipment to the DBP in the late 1970s, Siemens also came to accept the possibility of increased competition in the German market. Thus, there was a convergence of national government and industry preferences in all countries.

3.4 RACE AND THE COMMISSION'S CAMPAIGN FOR A CENTRAL ROLE

Continuing to seek a central role in policy development, the Commission sought to coordinate Research and Development in Advanced Communications Technologies in Europe (RACE) towards the end of 1993.[16] The SOGT, however, was unsupportive of the Commission's initial RACE proposal because the national PTTs continued to oppose further extension of EC competencies. Once again, the Commission turned to an epistemic community for assistance in its campaign to gain control over the sector. The ITTF mobilised the expert knowledge of 80 technical experts from the EC's 12 largest electronic companies and national research institutes (Esser and Noppe 1996, p. 556).

During the summer of 1984, this epistemic community produced a comprehensive plan for collaborative research in telecommunications technologies. Nevertheless, at the end of 1984, the SOGT rejected the Commission's proposed 1bn ECU RACE programme. Thus, although the Commission had begun by the early 1980s to establish itself as an important supranational actor in the European telecommunications sector, the national PTTs retained sufficient institutional authority to thwart certain Commission policy objectives.

Determined to extend its authority over the sector, the Commission nevertheless put forth a smaller 'Definition Phase' of RACE to a joint meeting of the telecommunications and industry ministers in June 1985.[17] The Commission proposal did not envision an important role for the national PTTs. The French, German and British ministers asserted, on the other hand, that the national PTTs should play a key role. More specifically, they argued that a reference model of the networks, terminals and services for Integrated Broadband Communications (IBC) should be designed within the CEPT (Sandholtz 1992b, pp. 242–244). French DGT Director-General Jacques Dondoux, for example, argued that the PTTs had to play a major role in RACE because they 'alone can ensure the interconnection of national networks' (Sandholtz 1992b, pp. 243–244). Against this, the Commission responded that the CEPT lacked the financial and organisational resources, contained too many non-EEC countries, and did not have the substantial relations with industry necessary for the successful implementation of the programme (Sandholtz 1992b, p. 244). Displaying the characteristics of a corporate actor, the Commission argued that the EC should control RACE.

In the end, however, the Commission had to bow to the demands of the French, German, and British governments (Sandholtz 1992b, p. 244). It was agreed that the CEPT would help define the RACE programme's activities through the preparation of the IBC reference model.[18] Thus, although the Commission was a successful policy entrepreneur in gaining Member State approval of its RACE proposal, the Member States nevertheless preferred to shift significant responsibility away from the EC and towards the intergovernmental policy-making arena of the CEPT. This is an interesting example of the futility of the Commission's attempt to exclude the PTTs from the policy-making process. As a strategy, it failed because the Commission was essentially seeking to exclude important and powerful policy stakeholders.

Hence, since this time, the Commission has pursued a revised strategy of including the PTTs within a broader policy network – within which the role of the PTTs and their influence has been incrementally eroded. For example, chapter six finds that, although the Member States have continued to demand an important role for the CEPT into the late 1990s, the intergovernmental CEPT has been forced to adapt its role as the Commission has continued to expand its authority in the sector. In this context, chapter six argues that the CEPT strategy has increasingly become one of damage control (that is, sharing of power).

Highlighting the fact that the CEPT had begun to adapt in accordance with developments at the EC level, it established a permanent liaison secretariat, the *Sécretariat pour les Spécifications et les Agréments* (SSA) to fulfil its obligation under the RACE preparation stage. This was the CEPT's first fixed administrative structure and was staffed by a permanent groups of experts (Permanent Nucleus) drawn from the national PTTs (Sandholtz 1992b, p. 244). From July 1985 to December 1986, approximately 400 industry experts from 109 private and public firms and organisations participated in the preparation phase (Commission 1987, p. 116). Industry handled the more technical work through the broadband group of the European Conference of Telecommunications and Electronics Industries (ECTEL). The more conceptually oriented aspects of network development, including the IBC network reference model, were handled by the *Groupe*

Spécial des Communications à Large Band-Permanent Nucleus (GSLB-PN), which was a CEPT broadband workgroup (Esser and Noppe 1996, pp. 556, 562).

Despite the determined efforts of the Commission, the intergovernmental CEPT thus maintained an important role in the RACE preparation stage. With industry urging their national governments to approve the plan along with the national PTTs, which realised they could maintain substantial influence over the plan, the Council approved the first phase of the RACE programme in December 1987.[19] This initial phase involved EC funding of ECU 550 million for five years for specific projects involving universities, companies or laboratories from a minimum of two EC countries.[20] Meanwhile, these recipients of EC funding became supporters of further Europeanisation of the policy sector,[21] thus further strengthening Commission legitimacy.

As a result of the development of RACE, the Commission emerged as a central actor in coordinating European research and development in telecommunications and as a hub of a supportive group network. The policy-making process became increasingly institutionalised at the EC level, while the policy network expanded, as a large and complex distribution of actors and organisations became drawn in. Sandholtz (1992a, 1998) argues that RACE was the EC's most important move into the telecommunications sector. The programme established the Commission as the:

> leader to which diverse liberalising interests would rally. From an analytical point of view, RACE is important because it was not the product of member-state initiative or direction. The Commission and its industry partners designed RACE and sold it to the states. (Sandholtz 1998, p. 150)

The Commission (1987, p. 5) noted that the greatest advantage of RACE was the 'climate of co-operation ... [and] establish(ing) mutual confidence between the public and private sectors in telecommunications'. RACE continued to strengthen the EC's basis of legitimacy in the sector, thus facilitating the introduction of more extensive efforts towards European restructuring.

The 1985 European Court of Justice (ECJ) ruling in the 'British Telecom case'[22] facilitated further CEC attempts to coordinate the restructuring of the European telecommunications sector. Consistent with institutionalist and neo-functionalist accounts of the ECJ as a key supranational actor with a bias for furthering European integration,[23] the Court ruled that the competition rules of the Treaty fully apply to the PTTs. The ECJ made clear that it would interpret monopoly rights narrowly and disfavour their extension to new services as technology developed. This decision was viewed by the Commission (1987, p. 124) as the 'cornerstone for the future interpretation of the Treaty with regard to telecommunications'.

Crucially, the 'British Telecom case' provided the justification for the consistent application of the Community's competition rules to telecommunications and helped to establish the Commission's regulatory authority within the sector. As will be seen in Part III of this book, the Commission seized upon EC competition rules as part of a regulatory strategy to coordinate the liberalisation of European telecommunications markets. The ECJ ruling in the British Telecom case, among other rulings discussed in chapter seven, has thus had a significant impact on the development of the European telecommunications sector.

The drive towards the Single European Market (SEM) provided a supportive policy frame for EU telecommunications liberalisation; it was a further important opportunity for the Commission to reinforce competitiveness in the telecommunications sector and, according to Competition Commissioner Karel van Miert, gave 'fresh significance to competition policy' (*Reuters European Community Report*, 18 May 1994). The European Council approved the Single European Act (SEA) in 1986, which committed the Member States to pursuing the completion of the SEM before 31 December 1992. The single telecommunications market was considered vital to the achievement of the SEM, because of its importance in ensuring the flow of information and supporting the development of trade, business and industry. As stated in the Commission's 1987 Green Paper on telecommunications (p. 2), 'the strengthening of European telecommunications has become one of the major conditions for ... achieving the completion of the Community-wide market for goods and services by 1992'. The SEA formally amended the Treaty of Rome and modified EC rules and procedures in order to increase the efficiency of Community decision making and facilitate the achievement of the SEM.

The SEA strengthened the Commission's authority in telecommunications through the extension of qualified majority voting (QMV).[24] This procedural modification transformed the climate of EC decision making and was essential in enabling the Community to progress towards creating a single market (Weiler 1997, p. 107). In addition, Title XV on Research and Development explicitly cited research and technology policy as a common European task, giving the EC formal competence over this sector for the first time (Peterson and Sharp 1998, p. 8). Explicit reference to a sector within the EC Treaty further legitimises EC action and increases Commission authority (Bulmer 1993 p. 365).[25] The power of the European Parliament was also extended by the SEA with the introduction of a second reading stage for legislation under the cooperation procedure.[26] Finally, the SEA conferred upon the Commission the powers to implement the rules that the Council lays down.[27] This extended the Commission's legal rights and widened its responsibility in the regulation of EC telecommunications initiatives.

In 1986, the foundation of the EC telecommunications field was consolidated through institution building and bureaucratic expansion; the ITTF merged with other organisational units of the Commission with jurisdiction over telecommunications, into an autonomous Directorate-General, DG for Telecommunications, Information, Industries and Innovation (DG XIII) (now DG Information Society). The Directorate-General's first major task was the production of the 1987 Green Paper *Towards a Dynamic European Economy – Green Paper on the Development of the Common Market for Telecommunications Services and Equipment*.[28] Described by Garfinkel (1994) as the first 'major threshold [event] in the evolution of European telecommunications markets',[29] the 1987 Green Paper proposed that the Commission play the leading role in coordinating the liberalisation and re-regulation of European telecommunications. As such, the entrepreneurial Commission sought to establish the EC as the primary institutional setting for the coordination of European telecommunications policies. In this context, the Commission was not merely arbitrating between competing interests, but was in the forefront of proposing innovative policy ideas that would upgrade the common interest and, perhaps, affect the character and substance of national cooperation in a more integrative fashion (see Pollack 1997, p. 122).

The 1987 Green Paper put forth several goals, such as establishing a set of harmonised access conditions to PTT networks, and initiated a process of policy construction, firmly institutionalised within the EC, to achieve these goals. The Green Paper thus shaped an iterative reform movement that produced most of the significant developments in the sector (Analysys Consultants 1994, p. 2).[30] The process has been iterative because technology is set out 'on a course of continuing development in which there are still many unknown factors' (Commission 1987, p. 4). As the Commission (1987, p. 6) explained,

> The proposed process is iterative; it accepts the existence of a movement, not all aspects of which can be defined today. The fundamental purpose of the measures is therefore to set off a dynamic process that will give the political, economic and social actors involved a better understanding of their own interests and to optimise their activities in the construction of the Community.

Consistent, once again, with both meta-theories of European integration, the 1987 Green Paper (Foreward) warned, 'national reform of the telecommunications sector in Europe will only be economically successful if it is tied into the larger Community context'. The Commission thus sought to Europeanise national policy making, as it has done in other sectors, through the coordination of national reform processes.

Every Member State had already begun to undertake an intense review of the telecommunications sector (Commission 1987, p. 24). The UK, for instance, had generally implemented the proposals under discussion, the Witte Commission in Germany, which had been meeting since 1985, was also considering such proposals, and such reforms had been planned by the Gaullist-led Chirac government in France since 1986. In addition, the CEPT had begun to review its organisation in response to, and connected with, the development of the EC's telecommunications policy (Commission 1987, p. 12). The situation was summarised by the Commission (1987, p. 2) as follows:

> The current wave of technical innovation resulting from the convergence of telecommunications and computer technology has now led to reviews in all Member States, and elsewhere, of the future organisation of the telecommunications sector and its necessary regulatory adjustment. The Commission considers it timely to aim at achieving maximum synergy between current developments and debates in the Member States, drawing fully on the potential offered by them to meet the objectives of the Treaty. This report is intended to launch a debate and to attract comment from a broad spectrum of opinion ... Ensuring that the varying national situations are fully taken into account in a European approach requires a wide-ranging debate over the whole of the Community.

The Commission thus sought to develop a pluralistic consultation process at the EC level, which would lead to: a common 'framing' (Rein and Schön 1991) of the problems faced; convergence towards its preferred, supranational solution; the generation of widespread support; and increased Commission legitimacy.

The 1987 Green Paper on telecommunications (p. 12) noted that the views of all parties concerned must be taken into account with respect to proposed regulatory changes. It sought to initiate 'a common thinking process regarding the fundamental adjustment of the institutional and regulatory conditions which the telecommunications sector now faces' (p. 10). It went on to state that the process should be focused on 'analysing common objectives and agreeing those means for achieving them which are best carried out in the Community framework [and] examining the external problems posed to all Member States by the rapid evolution of the world market' (p. 19). As such, in line with Kingdon's framework (1995), Commission officials continued to link their pet supranational solution of a coordinated EC approach, with the pressing problem that external pressures posed for the Member States.

The 1987 Green Paper on telecommunications (p. 179) aimed to take advantage of the consensus for reform in all Member States in developing its coordinated Community-wide approach: 'The current convergence of regulatory positions in the Member States provide a unique opportunity to synchronise current national moves and give them a European scale and dimension'. The Paper (p. 12) noted that 'there is a consensus in the Community that competition ... must be substantially expanded'. Nevertheless, in line with the assumptions of historical institutionalism, the Green Paper recognised the path dependent nature of the national PTT administrations. Thus, it recommended that changes be introduced gradually: 'Time must be allowed for present structures, which have grown up historically over a long period, to adjust to the new environment' (p. 12).

To serve as the basis of the consultation, the Commission put forth a two-pronged strategy to achieve comprehensive reform of the European telecommunications sector. The first prong was to liberalise the provision of telecommunications equipment and services (examined in Part III of this book). The second prong sought to re-regulate, or harmonise, national regulations at the European level (examined in Part II of this book). The Commission (1987, p. 65) argued that harmonisation measures were necessary to prevent 'a series of contentious cases and lengthy conflict (which would have to be resolved by the Commission under its competition authority)'.

Demonstrating the Commission's pluralistic consultation process, and consistent with Mazey and Richardson's (2001) argument that the Commission has 'played a crucial role in fostering the institutionalisation of interest intermediation in the EU', the Commission forwarded the consultative 1987 Green Paper on telecommunications to the Council; the European Parliament Economic and Social Committee (ECOSOC); the PTTs; telecommunications, data processing and services industries; user associations; unions; and several other social interest organisations (Commission 1987, Foreward). These organisations were then invited to submit opinions on the Green Paper and, by January 1988, more than 45 written submissions were received.

Clearly beneficial to Commission ambitions, a large, supportive issue network, encompassing a wide range of affected interests with unequal resources, was established at the EC level as a result of the consultation process on the 1987 Green Paper (Commission 1988a). The Commission consulted the SOGT and the national PTTs intensively (Commission 1988a; 1991). The PTTs and their associated organisations maintained a shared interest in the current system and therefore expressed concern over the proposals. However, they were in a minority.

The private sector, represented by several major industrial operators and various national, European, and international user organisations, believed that liberalisation would result in lower prices and better services and, therefore, were supportive of change (Commission 1988a; 1991). Unsurprisingly, these 'users' of telecommunications services responded enthusiastically to the two-pronged strategy outlined in the Green Paper because it was in their interest. Similarly, the Roundtable of European Industrialists (1986, pp. 16–19), whose membership includes many of Europe's largest multinational corporations, also endorsed the entire Green Paper. 'Among users', their report declared, 'the will exists to bring the EEC's objectives to fruition.'

The policy proposals of the employers' association, UNICE, whose membership includes equipment manufacturers and large users, echoed the Green Paper. Meanwhile, the International Telecommunications Users' Group (INTUG) strongly supported the Green Paper and argued that in some areas, such as reserving the sale of the first handset for the PTTs, the Green Paper should have gone further (McKendrick 1987). Finally, the US Chamber of Commerce, the Dutch Business Telecommunications Users Associations and Digital Equipment (DEC) all demanded that the exclusive rights over voice telephony services be abandoned. Thus, a broad coalition of support for Commission initiatives was mobilised.

Following the consultation process outlined above, the Commission fine-tuned the measures proposed in the 1987 Green Paper on telecommunications, in regard to both content and timing. In February 1988, a programme was submitted to the Council, the European Parliament and the Economic and Social Committee, for implementing the Green Paper (Commission 1988a). The Council subsequently passed, on 30 June 1988, a legally non-binding Resolution on the Development of the Common Market for Telecommunications Services and Equipment up to 1992 (Council of the European Communities 1988). Although non-binding, this resolution was an important high-level endorsement of Commission action; it sought to implement the objectives of the Green Paper through an action plan that aimed to achieve a balance between the twin objectives of harmonisation and liberalisation. The Council requested the preparation of proposals for Council directives for the rapid definition of harmonised technical conditions, usage conditions and tariff principles. The policy-making process surrounding these harmonised conditions will be examined in Part II of this book.

Member States, however, sharply disagreed on which particular services to include in the liberalisation programme. On the one hand, the UK, with the support of Denmark and the Netherlands, actively supported the extension of competition throughout the EC in all services, including basic voice services. These Member States either had liberalised already, or were in the process of liberalising, all telecommunications services. This prior liberalisation, therefore, served as the basis for their national positions in

EC negotiations. However, a majority of Member States, including France and Germany, wanted to reserve a number of services, including voice telephony, data transmission and telex, for the exclusive provision of their national PTTs.

The national positions of France and Germany, as explained in chapter two, were based on the fact that they had not undergone the same degree of liberalisation as the UK. National actors in these countries were therefore less prepared for competition in telecommunications services than their British counterparts. *Dirigiste* France, in particular, wanted to maintain national authority over the sector, while the deep-seated notion of *service publique* compelled French civil servants to support continued cross-subsidisation of services in order to ensure universal service. As a result, France sought to define 'reserved services' as widely as possible (Cawson *et al.* 1990, p. 195). Part III of this book examines the policy-making process surrounding the eventual liberalisation of telecommunications services and infrastructure.

3.5 CONCLUSION

This chapter has examined the construction of an institutional framework for the coordination of European telecommunications policies from 1957 to 1988. In line with the historical institutionalist assumption that institutions are not mere arbitrators but key players in policy development, the central argument defended is that the European Commission was an important policy entrepreneur during the period examined. Indeed, as highlighted above, these years were marked by continual efforts of the Commission to gain legitimacy in and, ultimately, authority over the sector.

Since the Treaty of Rome did not provide specific legitimacy in telecommunications, the Commission (prior to the SEA) needed the unanimous approval of the Member States in order to extend EC competence into the sector. Commission attempts in the 1960s and the early 1970s failed to overcome the determined opposition of the national PTTs. As rational actors, the PTTs sought to preserve their decision-making autonomy in the sector and to limit EC encroachment upon their work. Rational choice institutionalism highlighted the way in which the PTTs established an institutional framework in 1959, the CEPT, which was more favourable to their interests than the EC. The CEPT then served as the sole coordination mechanism for European telecommunications policies until the late 1970s, when the Commission began to take numerous small steps to develop an autonomous and active telecommunications policy.

During the late 1970s and 1980s, the Commission funded external studies to give credibility to its desired policy line; warned of the threat of US domination; and used other actors, such as industry and users, to help it gain authority, by drawing interest groups into the policy process. By these means, the Commission went from having no role to constructing and placing itself at the centre of an increasingly complex European policy network. The Commission, therefore, extended competence into a policy sector that had previously been the sole domain of the national PTTs through the intergovernmental CEPT.

Kingdon's framework (1995) is helpful in accounting for the timing of Commission policy development. The entry of Commissioner Etienne Davignon in 1977 marked the beginning of an EC telecommunications policy. Davignon's ability to mobilise supporters, create committees and institutionalise the policy-making process at the EC level was crucial to this period. Member State cooperation was organised and the foundation of a European policy network was laid. Davignon was able to link the increased technological and economic pressures discussed in chapter two, with the need for a coordinated European response. Meanwhile, the CEPT was forced to adapt its role as the Commission expanded authority into the sector.

The national positions of key actors in the UK, France and Germany, as discussed in the previous chapter, also helped account for the speed and timing of the developments uncovered. Prior to the early 1980s, the Member States were content with monopolistic national PTTs and, correspondingly, intergovernmental cooperation through the CEPT. However, increased technological and economic pressures, also examined in chapter two, prompted the emergence of reform processes in each Member State by the mid-1980s. This shift in national preferences was crucial. Consistent with intergovernmentalism, it was only then that the Commission was able to establish itself as a key actor in coordinating the reform of national regulatory frameworks. The role that national actors have continued to play in the EC policy-making process after 1988 and, in particular, national actors in the UK, France and Germany, will be examined in Parts II and III of this book.

Along with the considerations above, further implications of this chapter for the rest of the book are as follows. Consistent with historical institutionalism, which is helpful in organising an 'exercise in process-tracing' (Armstrong and Bulmer 1998, p. 56), the 1987 Green Paper on telecommunications established an iterative movement for the development of European telecommunications policies. This process is based on a dynamic reaction to technological and economic developments; widespread, pluralistic consultation with affected interests; the convergence of Member

States preferences; and the liberalisation and harmonisation of national regulatory frameworks. Thus, the patterns of policy-making uncovered in later chapters are often similar to those uncovered in this chapter. For example, chapter six demonstrates that the battle for authority between the CEPT and the EC through 2001 is rooted in the politics examined in this chapter. Finally, a transnational policy network of governmental and non-governmental actors and institutions was created at the EC level. Later chapters demonstrate that this policy network has continued to play an important role in the development of EC telecommunications policies.

The rest of this book examines the iterative policy-making processes established by the Commission's 1987 Green Paper on telecommunications. Part II analyses the policy-making process for the harmonisation of European telecommunications legislation (positive integration) and Part III analyses the policy-making process for the liberalisation of the EC telecommunications sector (negative integration). More specifically, these sections assess the extent to which distinctive patterns of policy making and policy outcomes (in terms of the ability of key actors to control or determine policy outcomes) have resulted from differences in institutional settings.

NOTES

1. See, for example, Downs (1967). There are also, undoubtedly, committed individuals such as Commissioner Etienne Davignon (see section 3.3) who become advocates for policy change because they believe in the underlying idea of liberalisation.
2. Directive 77/62/EEC, OJ L 13, 15.1.1977.
3. This pattern of expansion is similar to early Commission activism in other policy areas, including sex equality and the environment, where the Commission's aim has been to build up sympathetic constituencies of interest in favour of EC legislation. See, for example, Hildebrand (1992) and Mazey (1995). This also serves the Commission's interest in enhancing the perceived democratic legitimacy of Community activities.
4. See Mazey and Richardson (2001), for a discussion of the bureaucratisation of the European Commission and the incremental expansion of its sphere of competence.
5. Peterson and Bomberg (1999, p. 203) argue that these early attempts 'backfired'. Williams similarly criticised the Commission for being 'excessively preoccupied with an executive power which itself really presupposes a political unity' (*Ibid.* quoting Williams, R. (1973) *European Technology: the Politics of Collaboration*, London: Croom Helm, p. 139).
6. Interview, DG Information Society (DG XIII) official, Brussels, March 1999.
7. Interview, DG Information Society (DG XIII) official, Brussels, March 1999.
8. Interview, DG Information Society (DG XIII) official, Brussels, March 1999.
9. Interview, DG Information Society (DG XIII) official, Brussels, March 1999.
10. Interview, DG Information Society (DG XIII) official, Brussels, March 1999; see also Schneider and Werle (1990), pp. 90–91.
11. COM(80)422 Final.
12. *Agence Europe*, 29 December 1979, p. 9; Council of the European Communities (1984a); Council of the European Communities (1984b); see also Schneider and Werle (1990, p. 91); and Sandholtz (1992b, p. 227).
13. 'Communication from the Commission to the Council on Telecommunications: Lines of Action', COM (83) 573.

14. COM (84) 277 final.
15. Dang-Nguyen, G. (1986) 'A European telecommunications policy: Which instruments for which prospects'. Unpublished paper, ENST, Brest, p. 322, quoted in Schneider and Werle (1990), p. 93.
16. For a more detailed account of the development of the RACE programme, see Sandholtz (1992b), especially chapter eight on 'RACE: Making Connections' pp. 209–256.
17. COM (85)145; and COM (85)113.
18. This reference model was to serve as the basis for the RACE programme. The objective of RACE was defined as the 'Community-wide introduction of Integrated Broadband Communications (IBC) by 1995'. See Council (1985) (85/372/EEC), L210/24.
19. Council 88(28) EEC.
20. The EC and the participants each paid half the costs. See Ungerer and Costello (1989), pp. 153–154.
21. Interview, DG Information Society (DG XIII) official, Brussels, April 1999.
22. *Italy v. Commission*, Case 41/83, Judgement of March 20, 1985; 1985 ECR 873.
23. See, for example, Burley and Mattli (1993) and Dehousse (1998). These views are discussed further in chapter seven. See also Stone Sweet and Brunell (1998).
24. Article 205(2) of the EC Treaty (formerly Article 148(2)). The effect that QMV has had on telecommunications policy making will be examined in chapter five.
25. The Commission's authority in telecommunications was strengthened even more significantly after the ratification of the Maastricht Treaty in 1993, and will be discussed further in chapter eight.
26. Article 252 of the EC Treaty (formerly Article 189c). Chapter five argues that this institutional modification, along with other modifications brought about by the Maastricht Treaty, has enabled the EP to become a key actor within the EU telecommunications sector.
27. Article 202 of the EC Treaty (formerly Article 145). The composition of the committees overseeing the implementation of EC telecommunications legislation has provided a significant source of tension between the various institutions and, in particular, the European Parliament and Council. This 'comitology' dispute highlights the importance of institutional rules on policy outcomes; it has had a significant impact on the policy-making process of the European telecommunications sector and will be discussed further in chapter five.
28. COM(87)290 final.
29. Garfinkel argues that the Commission's 1992 Review of the Telecommunications Services Sector, as discussed in chapter seven, was the second watershed.
30. The iterative policy-making process established by the 1987 Green Paper is the subject of Parts II and III of this book.

PART II

The harmonisation of European telecommunications policies

4. The formulation of European Union telecommunications policies

INTRODUCTION

Part II of this book uses new institutionalism and the various actor-based approaches discussed in chapter one to analyse the re-regulation or harmonisation of telecommunications legislation at the EC level. This chapter analyses the sub-systemic level, or the policy formulation stage, of the EC telecommunications policy-making process. First, it highlights the way in which the Commission, as the result of its campaign to expand authority over the sector (see previous chapter), finally acquired a central policy-making role. Secondly, it explains and highlights the growth of a European interest group system linked to this policy sector. These two developments, combined, provided the basis for a more pluralistic policy-making environment.

This chapter focuses on the sub-systemic level of analysis, in a temporal sense, as the policy formulation stage until the Commission forwards a draft proposal to the Council of Ministers and the European Parliament. In so doing, this chapter relies on a stable, structural feature of the policy-making process to form a line between the sub-systemic and systemic levels of analysis.[1] These structural factors (that is, the EC Treaty rules and procedures governing the policy-making process) provide the Commission with formal agenda-setting authority and ensure that it is a central actor in the formulation of all EC legislation (Pollack, 1997). In this way, the sub-systemic policy making stage is the initial policy development phase, confined primarily to the Commission and informal networks. The policy networks, advocacy coalitions and epistemic communities approaches are the most useful analytical frameworks for examining this early stage of policy formulation.

This chapter comprises two substantive sections. The first section examines the establishment of a European institutional framework for the harmonisation of European telecommunications policies. It seeks to identify and explain the context for telecommunications policy making and highlights the way in which interest groups increasingly organised at the EC level. The second section examines the formulation of telecommunications proposals within the EC and evaluates the historical institutionalist hypothesis that institutions have an independent influence on policy outcomes. It discusses, in greater detail, the consultation process that shaped the proposals and highlights the impact of pluralistic consultation structures on the ability of users to influence proposals. In line with historical institutionalism, this chapter argues that particular institutional configurations affect policy formulation. In particular, the discussion shows how the institutional modifications of the early 1990s enabled the Commission to respond increasingly to user concerns in the formulation of telecommunications legislation.

4.1   PTT DOMINANCE OF EARLY EC TELECOMMUNICATIONS LEGISLATION

Continuing to act as a policy entrepreneur at the EC level, and following the strategy put forth in the 1987 Green Paper on telecommunications (see previous chapter), the Commission set out to re-regulate or harmonise the EC telecommunications sector. The Commission (1987, p. 65) sought to establish 'common principles regarding the general conditions for the provision of the network infrastructure by the telecommunications administrations to users and competitive service providers, in particular for trans-frontier provision'. These measures, termed Open Network Provision (ONP), were designed to harmonise access conditions for telecommunications networks and services throughout the EC. ONP aimed to establish a level playing field for new market entrants and facilitate the provision of new competitive telecommunications services.

Nonetheless, until 1990, when the institutional setting of the sub-systemic level was reorganised (see section 4.2 below), the national PTTs prevented the Commission from achieving the harmonisation goals put forth in the 1987 Green Paper. From 1987 to 1990, telecommunications legislation was formulated within a closed policy community dominated by national PTT representatives. As such, a very limited number of participants, with similar resources and a shared interest in maintaining the *status quo*, maintained a balance of power over policy formulation. Early telecommunications proposals thus reflected the dominance of the national PTTs in policy formulation, because they held a privileged position in the policy-formulation process through the SOGT advisory committee. As Analysys consultants noted in a self-funded study:

The process for specifying ONP is flawed: The [PTTs] dominate the process that is producing the conditions under which ONP will be provided, and are thus defining the conditions under which their future private competitors will operate. The [Commission] has reinforced the position of the [PTTs] by making a [PTT]-dominated body its advisor on technical and economic aspects of ONP. (Analysys Consultants 1989, pp. xi–xii)

A UK user group representative to the SOGT committee from 1987 until 1990 noted that being in an SOGT meeting was 'like being in a CEPT meeting. The [PTTs] were dominant, very retractive, aggressively defending the need not to change'.[2] Significantly, the PTTs, and not the Commission, chaired the SOGT meetings and controlled the early development of telecommunications proposals.[3]

In line with historical institutionalism, and as discussed in chapter three, Davignon had created the SOGT in 1983 at a time when the Commission needed to bring the national PTTs into the EC policy-making process in order to expand the Commission's own legitimacy in the sector. Since all EC Member States maintained monopolistic PTTs into the 1990s (with the exception of the UK and Germany),[4] Commission officials recognised that the PTTs had to be included in the policy community, and have influence in policy formulation, if the Commission was to gain Member State approval (in the Council of Ministers) of its initiatives.[5] The national PTTs, therefore, maintained a foothold at the European level through the SOGT, which controlled the formulation of EC telecommunications legislation from 1987 to 1990. As a result of this PTT dominance during the early years of the policy-formulation process, early ONP proposals primarily reflected the interests of the PTTs (in limiting access to their networks) and not the interests of those seeking access to PTT networks (that is, users).

Since the national PTTs formed a restricted policy community until 1990 in policy formulation, users had limited influence over early proposals. The term 'users' is used throughout this chapter to refer to users of the public telecommunications network; this includes end users (final users of telecommunications services or consumers) such as business and household subscribers as well as companies wishing to provide services over the network (service providers).[6] Since the 1987 Green Paper on telecommunications had sought to establish an *'agreed* set of conditions for open network provision (ONP) *to service providers and users'* (italics added), they might have been expected to have had a key role in negotiating access conditions to the public network. As 'the main beneficiaries of the new opportunities', the 1987 Green Paper (p. 2) specifically recognised the importance of including users in the consultation process. The European Council of Telecommunications Users Associations (ECTUA 1987) also recognised the need to include the 'full participation of user representatives'.

Similarly, Commission officials recognised the value of user input in developing access conditions and supporting the Commission position over the resistance faced by recalcitrant PTTs. According to one of the Commission officials in charge of the early policy-making process, user input was viewed by the Commission as essential justification for the Commission's attempts at coordinated reform.[7] Commission officials thus viewed user input as valuable, not only for their expertise in developing feasible access conditions to PTT networks, but also for gaining further legitimacy over the sector. Seeking to expand the consultation process to include a more prominent role for users is a classic example of how the Commission sought to expand its own role in the sector.

Nevertheless, although user input into the early policy-making process was considered important by both users and Commission officials, the only formal, organised contact that user group representatives and service providers had with early proposals was through Commission workshops, which were designed to inform users about draft proposals and solicit their reaction (Austin 1994, p. 102). These workshops, however, did not involve users significantly in the development of the proposals. As a self-funded study by Analysys Consultants (1989, p. xii) concluded, 'there is no evidence that the opportunities for public consultation on ONP proposals have had an effect on the process ... users are being alienated by the [PTT] dominance in the setting of the ONP conditions'. Although officials recognised the value of user input, the Commission's institutional framework privileged the national PTTs, thus enabling them to maintain a closed policy community at the European level through the SOGT committee. As a result, user influence was limited during this early period of policy making.

A sub-group of the SOGT, the *Group d'Analyse et de Prévision* (GAP), was directly responsible for the formulation of early ONP proposals. GAP met five or six times a year, from 1987 to 1990, to negotiate the development of ONP principles. Member States were free to select participants for their national delegations. Only the UK delegation, however, included user representatives. Reflecting the broadening of their policy network (as examined in chapter two), the UK delegation included national trade and user groups, along with the private interests of BT and Mercury and their national regulatory authority, OFTEL.[8] France and Germany, as explained in chapter two, had not pursued extensive regulatory reforms in the late 1980s and therefore their national delegations were composed of national PTT representatives. Because, at this time, the national PTT also served as their governmental ministry and regulator, the stance of French and German governments clearly lay with the interests of their national PTTs.

Because of the dominance of the national PTTs in the early formation of ONP principles, the Commission was heavily criticised for a lack of openness in the policy-formulation process. European and International user groups, such as the European Council of Telecommunications Users Associations (ECTUA), claimed that their members were not sufficiently involved. Other user groups such as the European Association of Information Services (EAIS), the International Chamber of Commerce (ICC) and the International Telecommunication Users Group (INTUG) also argued that they were less than fully involved (Analysys Consultants 1991, p. 39; Mansell 1993, p. 74). INTUG (1989) argued that:

> The ONP analysis work is being carried out by the GAP, an EEC group mainly composed of representatives of the [PTTs]. Thus, the work is proceeding mainly from a [PTT]-oriented perspective ... Instead of proceeding from the premise that ONP is intended for users, [PTT] involvement in the process has diverted the initiative into what [PTTs] are willing to make available (in terms of access), within a deliberate time frame.

In its self-funded study, Analysys Consultants (1989, p. 69) argued, 'the reliance of the Commission on the SOGT will continue to cause imbalance as long as the composition of the national delegations to the SOGT remains unchanged. It is difficult to see how this situation can continue in the long term if ONP is to develop seriously.' Path dependency, therefore, played an important role in the development of European telecommunications policies until the 1990s. The Commission and the dense network of powerful national PTTs continued to rely on the SOGT and its GAP subcommittee as coordination mechanisms. PTT influence thus remained locked in at the European level.

Institutions clearly structured the access of actors to the policy-making process and shaped policy outcomes. As a result of their privileged institutional position, early proposals largely reflected the views of the national PTTs. For the GAP, ONP constituted an important market opportunity for the PTTs to provide new services, while allowing new service providers access to their networks on their terms. An initial GAP (1988) study on the development of ONP principles stated that:

> It is envisaged that ONP could represent a new range of commercial offerings by the [PTTs] ... The usage conditions and tariff arrangements that apply to ONP offerings should be such as to make them attractive to Private Services Operators, taking account of [PTT's] other operational and commercial constraints and obligations ... However, since Open Network Provision is seen as a natural evolution of the current offerings of the [PTT's], existing technical and operational functions will be adopted whenever appropriate.

GAP thus intended ONP to advance the development of non-reserved services, increase competition between the PTTs and new service providers in those services, and increase the usage of the public network. GAP was aware that ONP might affect the marketing and planning of the PTTs. GAP, however, did not intend ONP to result in a loss of PTT market share. The Commission largely based its first ONP proposal in January 1988 on this GAP study. Predictably, INTUG noted that this initial proposal 'tend(ed) to reinforce the position of the [PTTs] ... [and was] weak in the promotion of fair competition between the [PTTs] and private service providers'.[9]

Beginning in 1989, the Commission began to recognise the degree to which the institutional setting was restricting user input into the policy formulation process and inhibiting achievement of the harmonisation goals set out in the 1987 Green Paper.[10] Herbert Ungerer, Director of the Commission's telecommunications policy division, admitted,

> we have been criticized during the whole of last year [1988], particularly on ONP, that the process was not open enough. I believe that part of the criticism was certainly justified. It is not entirely our responsibility ... we are working with committees, and the composition of the delegation in those committees are clearly defined by member states – the Commission talks to member states. (quoted in Analysys Consultants 1989, p. 62).

Although the Commission recognised and acknowledged the lack of openness, it sought to deflect criticism from itself, by shifting some responsibility for the structure of the policy-making process to the Member States. Nevertheless, following this recognition, and at least partly resulting from the Commission's own interest in opening the process in order to expand its own authority in the sector, and the mounting exogenous pressures for reform (see chapter two), institutional adjustment to the policy formulation process took place in the early 1990s.

4.2   THE OPENING OF THE POLICY FORMULATION PROCESS

'Fundamental changes really only began with the very, very important step of the development and agreement of the Services Directive and Framework Directive which led to the setting up of the ONP Committee.'[11] Taking advantage of the reform processes occurring in each Member State during the late 1980s (see chapter three), the Commission used its substantial competition authority under Article 86 of the EC Treaty (formerly Article 90) to adopt a Services Directive liberalising all value-added telecommunications services on 28 June 1990 (see chapter seven).[12] Designed to create a more market-oriented regulatory environment, the Services Directive (Article 7) mandated a separation between the operational and regulatory functions of the national PTTs. Although this separation had occurred already in the UK in the early 1980s, all other Member States were required to introduce similar organisational changes prior to July 1991. Thus, in each Member State, a national telecommunications operator (TO) retained control over the public telecommunications network and services, while a separate national regulatory authority (NRA) was established to perform regulatory functions.

Based on the Community's normal legislative procedures under Article 95 of the EC treaty (formerly Article 100a), the Framework Directive on the application of ONP principles to the telecommunications sector was adopted by the Council of Ministers on the same date, 28 June 1990.[13] The ONP Framework Directive (Article 9) established an advisory committee of Member State representatives, nominated by COREPER officials, to assist the Commission in the development of ONP conditions. The Communications Committee (formerly the ONP Committee) has met about seven or eight times per year since it was established in 1990.[14] In terms of composition and the Commission's role, this Committee marked an important departure from the earlier institutional balance of power. Rather than being made up of PTT representatives, which was the case with the SOGT and GAP Committees discussed above, the Communications Committee comprises mostly NRA representatives. Thus, national representatives on the Communications Committee are generally more supportive of the Commission's initiatives. In contrast to the GAP meetings, which were chaired by a PTT representative, the Communications Committee meetings are presided over by a Commission official. Thus, a more liberal policy network was institutionalised at the European level to replace the less liberal PTT policy community.[15]

Meanwhile, the Commission also sought to institutionalise further interest representation at the European level. The ONP Framework Directive (Article 9) required the Communications Committee (ONP Committee) to 'consult the representatives of the telecommunications organizations, the users, the consumers, the manufacturers and the service providers'. Thus, the TOs (former PTTs) were henceforth to be only one group among many with which the Commission and the Communications Committee consulted. Realising that it would strengthen user policy influence to form a single forum, users lobbied Michel Carpentier, the Director-General responsible for telecommunications, to assist in the formation of a consultation forum.[16] Users contended that it would be more effective and efficient to have a single focus for discussions with the Communications Committee (rather than having the Communications Committee consult with each of these groups individually).[17]

In response, the Commission released a document in December 1990 calling for the creation of a platform for interests to be represented to the Communications Committee (ONP Committee). As the Commission (1990a, pp. 1, 4) noted,

> To ensure that this consultation is truly representative of the views of all interest groups there must be equal opportunity to input to the process and transparency in the operation of the process. This opportunity must be available to all members of relevant interest groups by a mechanism that does not favour one interest group or another ... The setting up of a permanent consultative body should therefore be seen as a means to provide a forum for the exchange of views within and amongst interest groups. In this way disputes between interest groups would be avoided by focusing the discussion and reducing the opportunity for misunderstanding.

The Commission thus sought to institutionalise a pluralistic consultative mechanism that would bring together interest groups, facilitate discussion between them and create an 'advocacy coalition' (Sabatier, 1987). This strategy was used by the Commission (as in other sectors) to gain support for its proposals and, more generally, to legitimise its intervention in the policy sector. Also, these groups provided the Commission with technical expertise.

Commission officials hoped this forum would lead affected interests to accept a common framing of proposals for the application of ONP principles to the telecommunications sector. As Mazey and Richardson (2001a; see also 2001) have argued, this is a way of depoliticising issues within the Commission to achieve the support of competing interests, which is essential to successful policy formulation:

> The Commission has played a crucial role in fostering the institutionalisation of interest intermediation in the EU ... Wherever policy is initiated, it is Commission officials who are charged with the task of drafting legislative proposals that are acceptable to affected interests and governments within (and beyond) the Union ... Not only do [affected interests] provide Commission officials with technical information and advice, the support of cross-national advocacy coalitions of groups is essential to the successful introduction of Commission proposals (Sabatier 1987). These functional incentives to consult groups are buttressed by the Commission's political needs to be able to demonstrate openness and thus enhance its own legitimacy.

The Commission's proposed structure for interest consultation is illustrated in Figure 4.1.

In February 1991, the European Telecommunications Platform (ETP) (formerly the ONP Consultation and Coordination Platform) was formed, comprising 30 representatives from the telecommunications industry. The ETP is self-financed through membership fees and is entirely independent

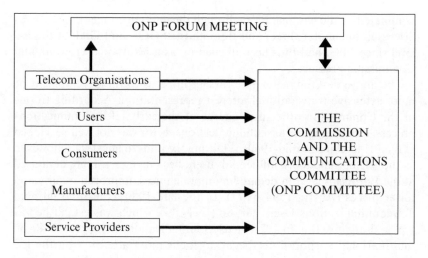

Source: Adapted from Commission (1990a), p. 2.

Figure 4.1   The institutionalisation of interest consultation

of the Community and NRAs.[18] Through a policy of open membership, the ETP solicits the views of all affected interest groups. The main objective of the ETP is to narrow and focus discussion on telecommunications issues through open discussion of the views of all affected groups.

> The principal function of the Platform will be to provide efficient and transparent mechanisms for the exchange of views among interest groups
>
> – to assist members to understand each other's position
> – to focus the discussion and to identify common views and areas of disagreement, and
> – to assist members to submit their views, or sets of common views, to the Commission, other institutions of the European Communities, and the [Communications Committee].[19]

The ETP thus institutionalised a new, more liberal policy network, centred on the Commission, in the policy-formulation process. New interests, including users, consumers, manufacturers and service providers, were brought into the policy-making process, further eroding the influence of the (former) PTTs. The ETP works more closely with the Communications Committee than the ONP User Hearings had worked with the SOGT and GAP.[20] In so doing, these interest groups are more closely linked to the development of telecommunications policies. The old, closed, policy community at the European level, which had previously been sustained and dominated by the national PTTs through the intergovernmental CEPT (as discussed in chapter three) and through the SOGT and GAP at the EC level since 1983, had thus been opened up as a result of user group and Commission pressure.

The above developments gave further impetus to the development of more extensive transnational interest representation. According to one of the Commission officials in charge of the early telecommunications process, lobbying in the telecommunications sector was not well developed at the EC level in the mid-1980s.[21] During the mid- to late-1980s, however, as the EC became more influential, users became increasingly organised at the European level in demanding more choices, improved quality and lower prices (Austin 1994, p. 97). In the late 1980s, the International Telecommunications Users' Group (INTUG), which was established as a global organisation in 1974 by telecommunications managers of 40–50 multinational corporations, began to focus more attention on the EC and actively lobbied the Commission to take steps to liberalise European telecommunications.[22]

Another user group, the European Council of Telecommunications Users Associations (ECTUA), was formed in 1987 to represent national user associations in the UK, France, Germany, Belgium, the Netherlands and Spain. The founders of ECTUA felt that it would be more effective and powerful to have an entirely European organisation representing European users to the Commission.[23] In particular, the ECTUA founders viewed INTUG as being tied too closely to American interests.[24]

Predictably, interest groups representing telecommunications users shifted their organisational resources and lobbying efforts to the EC level as power shifted away from national capitals. At the same time, to bolster its own position by increasing awareness of (and support for) its policy proposals, the Commission supported the development of this EC level network of interests. As Mazey and Richardson (2001a) point out, there is often a 'symbiotic relationship between the Commission and groups. Groups are, of course, drawn to Brussels by a desire to defend and promote the interests of their members in the context of EU policy-making. The Commission is however, equally dependent upon groups.' For example, the Commission provided meeting rooms and assistance with travel expenses to a Round Table on ONP established by ECTUA (Austin 1994, p. 102). Commission officials assert that this Round Table provided an important avenue of user support for, and user expertise into, Commission proposals during the early 1990s.[25] Thus, by 1990, a European interest group system linked to this policy sector had been institutionalised at the EC level.

Meanwhile, the CEPT and the national TOs were forced to adapt once again as the Commission further extended its own responsibility over the sector. Having lost their privileged position in the ONP policy-making process, the 26 former PTTs of the CEPT recognised the need to establish their own European lobby. They thus formed the European Association of Telecommunications Network Operators (ETNO) in 1991 to lobby the Community institutions. ETNO took over some CEPT responsibilities, became the general policy body for the TOs and began to develop and defend common positions at the EC level. More specifically, ETNO sought to oppose the Commission's authority with respect to telecommunications harmonisation. The first ETNO President, Klaus Grewlich, sought to enforce the operators' role in the process of reform:

> While it is incumbent upon the governments and the European Commission to promote the liberalization process and to ensure that privatization of the sector develops in an orderly way, it is not the Commission's responsibility to design future networks and to choose standards. This is the task of operators. (quoted in *Agence Europe*, 26 May 1993)

The formation of ETNO perhaps most clearly represents the extent to which policy-making power had shifted to the EC level. Whereas in the mid-1980s, the Commission was still seeking to gain legitimacy in the telecommunications sector, by the early 1990s, it was the former PTTs, which, having previously monopolised European level coordination, were struggling to retain their, by now, limited influence. The EC had become the dominant institutional arena for the coordination of European telecommunications policies. Figure 4.2 demonstrates the lines of communication

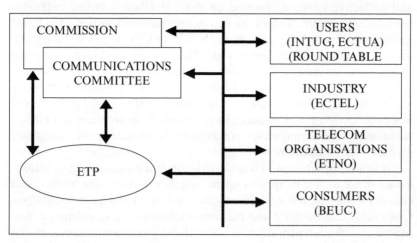

Source: Adapted from Commission (1990a), p. 4.

Figure 4.2   A European telecommunications policy network

that were established between these newly formed interest groups and the Commission.

Both governmental and non-governmental officials viewed ETNO during the early 1990s as a 'club' for the former PTTs, which sought to defend the interests of the former monopolists. As such, its views were opposed to the policy objectives of the Commission. ETNO's external relations manager and spokesman commented that, 'putting it very simply and crudely the 1998 package was directed against the operators who were then the members of ETNO. So, it was not at all surprising that sometimes ETNO said "It is terrible!" and the Commission said "Yeah, well it's designed to be terrible!"'[26]

The 1990 ONP Framework Directive (Commission 1990) discussed above also required the Commission to publish initial draft proposals and allow a three-month public comment period before making final draft proposals. According to Article 4(4), the Commission shall 'invite ... public comment by all parties concerned on the reports on the detailed analysis ... The period for submitting such comment shall be not less than three months.' This requirement further institutionalised the representation of interests at the EC level. Although GAP proposals had generally been made available for public comment, many users saw the codification of the three-month public consultation process in the Framework Directive as an extremely important step towards making the policy-making process more transparent and more democratic.[27]

The mandatory consultation period enabled the individual opinions and collective positions formed on draft legislation during extensive consultations within the ETP, as well as the common positions developed within other organisations such as INTUG, ECTUA and ETNO, to be expressed publicly to the Commission. Due to their previous consultation within these organisations, affected interests were able to put forth more cohesive, informed and reasoned opinions. Commission officials contend that these more refined opinions have increased the impact that affected interests have had on Commission proposals.[28] In an effort to increase transparency, the Commission has published the submitted public comments on draft ONP legislation.

Historical institutionalism is useful in highlighting changes in institutional values that have a subsequent impact on policy making. The institutional changes discussed above, including the creation of the Communications Committee and the ETP and the introduction of the mandatory three-month consultation period, are in line with the growing importance placed on openness and transparency throughout the European Communities since the early 1990s. The Community's normative change is apparent in the declaration on transparency issued at Maastricht in December 1991, which stated, 'the Conference considers that transparency of the decision-making process strengthens the democratic nature of the institutions and the public's confidence in the administration' (quoted in Lodge 1994, p. 349). The High Level Group on the Operation of the Internal Market reported to the Commission in 1992 that 'wide and effective consultation on Commission proposals is essential'.[29] The issue was then discussed at the 1992 Birmingham and Edinburgh summits, which led to two December 1992 Commission communications and a series of 'openness initiatives' in 1993 and 1994.[30] In this way, as historical institutionalism claims, it is difficult to separate the formal rules and procedures of governance from the norms and codes under which policy makers operate. Indeed, norms are part of the institutional context under this framework. As the norm of openness and the culture of consultation strengthened (because of the need to gain public support for integration), the advocacy coalition in favour of liberalisation was strengthened and the TOs (former PTTs) were weakened.

By the early 1990s, therefore, the closed policy community for the coordination of European telecommunications policies, which had previously been dominated by the national PTTs, had been opened. At the same time, the policy-formulation process and interest consultation had become highly structured, and a powerful advocacy coalition had been established at the ÉC level. As such, the policy networks approach is less capable of accounting fully for, and explaining the totality of, policy formulation at the sub-systemic level. As applied below, historical institutionalism has correspondingly become more relevant. An iterative process was developed for the application of ONP principles to various aspects of the telecommunications sector. The policy-formulation process since the early 1990s is illustrated in Figure 4.3 and Table 4.1.

Table 4.1   Ten steps for the formulation of telecommunications proposals

STEP 1	Identification of area to be studied by Commission
STEP 2	Independent study commissioned
STEP 3	Final study report to Commission
STEP 4	Public presentation of study results
STEP 5	Publication of Analysis report by Commission
STEP 6	Public consultation period
STEP 7	Draft proposal for a COM Directive
STEP 8	Discussion in Communications Committee
STEP 9	Proposal adopted by Commission
STEP 10	Proposal forwarded to Council and EP

Source: Adapted from Figure 2.1 in Analysys Consultants (1992), p. 10.

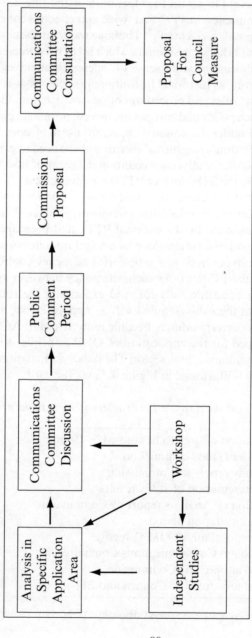

Source: Adapted from Commission (1991f), p. 2.

Figure 4.3   Procedure for the formulation of telecommunications proposals

Indicative of the highly structured nature of the sub-systemic level, both the Commission and Analysys consultants have identified several distinct procedures or steps in the process of policy formulation in their attempts to describe and explain the development of proposals.[31] The rest of this chapter evaluates the extent to which the pluralistic consultation procedures identified in these steps have affected policy formulation, through a case study on the application of ONP principles to voice telephony.

Voice telephony is chosen as the case study because it is the largest and most profitable area in the telecommunications sector, while the ONP Voice Telephony Directive is an important component of the EC regulatory framework. The case study assesses the ability of historical institutionalism and actor-based approaches to account for policy formulation at the sub-systemic level. It is argued that historical institutionalism has become a more relevant explanatory approach as the policy-making environment has been increasingly institutionalised. That is, as institutions increasingly structure the early stages of policy formulation, analysis of the role of institutions in policy formulation likewise becomes more important. Since there are indications that this process of institutionalisation is occurring in other policy sectors (Mazey and Richardson 2001, 2001a), new institutionalism is likely to become an increasingly relevant framework for explaining policy formulation throughout the EC's sub-systemic level.

Clearly, not all elements of telecommunications proposals are formulated within the ten steps identified in Table 4.1. Much lobbying and consultation is unofficial, unrecorded and takes place outside of formal consultative mechanisms. In addition, not every procedure necessarily occurs in the precise order outlined in Table 4.1. For example, the Commission often publishes an initial proposal (step 5) before the final report from external consultants (step 3) is available. Nonetheless, in seeking to test historical institutionalism, the rest of this chapter assesses whether or not these procedures have had an important and influential effect on policy formulation.

4.3   THE FORMULATION OF THE ONP VOICE TELEPHONY DIRECTIVE

This case study assesses the extent to which the procedures or steps of the policy formulation process, identified by both the Commission and Analysys Consultants (see Figure 4.3 and Table 4.1), have been institutionalised at the EC level and whether they influence the formulation of telecommunications legislation. This detailed examination of the sub-systemic level of the policy-making process enables the testing of historical institutionalism and the application of the actor-based approaches examined in chapter one. It finds that this highly institutionalised setting provides for a more pluralistic policy formulation process than the pre-1990 policy formulation process examined in section 4.1.

As PTT prominence in the post-1990 institutional setting was reduced, the Commission, independent experts and users gained more powerful positions in policy formulation. In contrast to several of the pre-1990 proposals, which were produced largely as a result of internal discussions within the GAP and SOGT committees, the Commission has consistently relied on external consultants for guidance in formulating later policy proposals. The Commission mobilised an epistemic network of experts with a recognised competence in, and technical knowledge of, voice telephony services. As Haas (1992, p. 15) points out:

> epistemic communities can help formulate policies ... [by] working out the details of the policy, helping decision makers anticipate the conflicts of interest that would emerge with respect to particular points, and then building coalitions in support of the policy. If the policy is instituted and problems ensue, the decision makers have the option of pointing to the information given to them by experts and spreading the blame.

Commission officials commissioned several studies on various aspects of voice telephony services, including a study by Margrit Sessions of Logica on intelligent voice services and by Wim van Dijk of the European Telecommunications Consultancy Organization (ETCO) on tariff principles.[32] The most influential and comprehensive study commissioned by the Commission, however, was a report on ONP principles and voice telephony produced by a 13-person team from National Economic Research Associates (NERA) in London.[33] The Commission thus employed at least 15 independent experts in the earliest formulation stage of the voice telephony proposal.

As the Commission further institutionalised interest representation at the EC level, the initial policy formulation stage became increasingly pluralistic. In February 1991, the Commission held a one-day workshop in Brussels to obtain public reaction to the initial findings of the consultants' studies. The 60 workshop participants included representatives of the users, consumers, manufacturers, service providers and TOs. A second one-day workshop was held in Brussels in September 1991, following the completion of the consultants' studies, which attracted more than 100 participants. Open workshop discussions, in which the reports from external consultants are presented to affected interest groups, have become a standard Commission procedure. User representatives view these workshops as an extremely important part of the overall consultation process, even though they are not part of the formal three-month consultation requirement stipulated in the Framework Directive.[34] Richardson (1996, p. 18) notes that:

> Commission officials are moving towards institutionalised structures which ... bring together groups of policy actors (be they epistemic communities, advocacy coalitions, or different policy communities) in a forum ... policy-makers are intent on securing agreement and stability, and recognise that this process must involve the participation of the various types of 'stakeholders' in the policy area or policy sub-system.

These workshops thus add another access point for interest groups to contribute to the pluralistic consultation process and have served to further institutionalise interest representation at the EC level. This increasingly large, loosely organised and varied set of interests involved in policy formulation at the EC level thus resembles an issue network rather than the closed policy community which controlled the process prior to 1990.

The epistemic community and issue network discussed above were influential in shaping the initial voice telephony proposal. In particular, the findings of the consultants' studies (including NERA progress reports, interim reports and a draft final report), meetings between Commission officials and the consultants, and the views expressed by industry and interest group experts were used by a team of five Commission officials in the production of an *Analysis report on the application of ONP principles to voice telephony* (Analysis report) in June 1991.[35] Following preliminary comments made by the NRAs in an early July 1991 Communications Committee meeting, the Commission officials modified the Analysis report and subsequently made it available for public comment in late July 1991 through a notice in the *Official Journal* (Commission 1991b, p. 12).

The Analysis report established an initial 'framing' (Rein and Schön 1991) of the policy problems and possible solutions in order to shape the subsequent debate and facilitate the consultation process. It addressed the main requirements for EC voice telephony services and put forward key proposals. Its three main goals were to:

- to establish the rights of users when dealing with the national TOs;
- to give competitive service providers access to the public network infrastructure on an equitable and non-discriminatory basis; and
- to meet the requirements of the single market, particularly through facilitating the provision of pan-European telecommunications services.

The remainder of this chapter examines the public consultation period and the Communications Committee discussions over the draft proposal in order to assess the extent to which the prevailing rules and procedures governing policy formulation had an important effect on policy development. In particular, the effect of the institutional modifications, discussed previously in section 4.2, on the ability of users, along with the TOs and the NRAs, to influence the formulation of telecommunications legislation will be assessed empirically. An examination of the policy-making process surrounding the public consultation period in section 4.3 (a) will be followed by an examination of the policy-making process surrounding the Communications Committee in section 4.3 (b).

4.3(a)   Increased Openness in Policy Formulation

The institutionalisation of a three-month public consultation period for all proposals provided for a more pluralistic policy formulation process, which enabled a wide range of interests to influence the voice telephony proposal. Following *Official Journal* notification in late July 1991, the public was given a three-month period, from August to October 1991, to comment on the Analysis report.[36] Comments were submitted to the Commission from thirteen TOs, nine user groups, seven service providers, five equipment manufacturers, three consumer associations, three TO associations and nine others from the public and private sectors (Austin 1994, p. 103).

The diverse group of affected interests that responded to the public consultation shares several of the characteristics of Marsh and Rhodes' (1992, p. 251) issue network: a large number of actors, wide range of affected interests, conflict ever present, and a consultative relationship. Nevertheless, it is perhaps even more analytically useful to separate this large group of actors into two smaller, competing advocacy coalitions (Sabatier, 1987). An advocacy coalition is composed of 'people from a variety of positions ... who share a particular belief system ... and who show a non-trivial degree of coordinated activity over time' (p. 660).

On the one hand, the users of telecommunications services (including user groups, consumer organisations and private service providers) believed in the virtues of a competitive telecommunications market. They felt that liberalisation and harmonisation would lead to lower prices and more advanced and better telecommunications services. They generally called for the strengthening of the provisions of the Analysis report (Commission 1991c, p. 5). Some users even called for the inclusion of additional features, such as the establishment of a European monitoring body (Commission 1991e).

The TOs, on the other hand, generally viewed the retention of certain monopoly rights as a requirement for a functional universal service. Although not entirely opposed to the Commission's attempt to apply ONP principles to voice telephony, the TOs predictably demonstrated concern at the possible introduction of over-regulation (Commission 1991c, p. 5). ETNO argued that certain provisions were too detailed, while other provisions fell outside the scope of the ONP Framework Directive (ETNO 1992).

It might also be argued that these groups were not united by shared belief systems, but by shared interests. Users clearly perceived their collective self-interest to lie in a liberalised European telecommunications sector while the TOs maintained a shared interest in preserving the *status quo*. As such, it is possible to conclude that 'coalition stability could be the result, not of stable beliefs, but rather of stable economic/ organizational interests' (Sabatier 1987, p. 663). Sabatier, however, points out that this distinction 'raises very thorny methodological issues, in part because belief systems are normally highly correlated with self-interest' (*Ibid.*). Setting apart these difficult methodological issues for the moment (and whether the separate coalitions of users, on the one hand, and TOs, on the other, were united by deep core beliefs, shared interests or (more likely) both), it is argued below that Sabatier's advocacy coalition approach is a valuable tool for analysing the formulation of telecommunications policies.

Immediately after the public consultation period, in November 1991, the Commission redrafted the Analysis report into the first *Draft proposal for a Council Directive on the application of ONP to voice telephony* (step 7).[37] In redrafting the proposal, the Commission served as a 'broker' (Mazey and Richardson 2001), whose main concern was to find a reasonable solution that would reduce political conflict between the competing advocacy coalitions (Sabatier 1998, pp. 652, 662). As the Commission (1991c, p. 5) explained, 'Balancing these views, the Commission took the view that over-regulation had to be avoided and, while appreciating the calls for further provisions, decided not to extend the proposed directive beyond the minimum provisions originally contained in the *Analysis report*' (italics in original).

An examination of key issues raised in the public consultation period demonstrates that the Commission acted as a policy broker, while users, in addition to the TOs, were able to influence the redrafting of the proposal. Six key issues raised during the public consultation period were analysed.[38] Confirming that the policy formulation process is more favourable to users, the modifications made by the Commission in redrafting the Analysis report into the November 1991 proposal were in line with the comments expressed by users on a majority (four of six) of the issues examined. These issues concerned the scope of the definition of 'users'; compulsory compensation for a TO's failure to meet contracted service quality; tariff transparency and cost-orientation; and accounting consistency and transparency. One issue, which concerned interconnection arrangements between competing networks and was the most highly disputed issue during the public consultation period, was modified in line with comments put forth by both advocacy coalitions. On the other hand, with regard to the provision of advanced services, the Analysis report was modified in line with the comments expressed by the TOs.

Although it is not possible to conclude with absolute certainty that individual comments made during the consultation period led to specific textual revisions, the Commission officials that drafted the November 1991 proposal contend,[39] and the proposal itself acknowledges (Commission 1991c, p. 5), that it took 'into account the comments' received. Marc Austin, who worked with the Commission as an independent expert during the early 1990s, produced his own study on the voice telephony consultation period (1994, p. 107), which concluded, 'it is hard to argue that users have been completely ignored by the Community policy-making process'. Marc Austin's own analysis (1994, p. 107) found that 'users appear to have won four and perhaps five of the seven issues raised by the Voice Telephony Directive, were handed a draw on at least one, and clearly lost one'.

The modifications to the Analysis report, at least partly, reflect the Commission's responsiveness to the comments expressed during the August to October 1991 public consultation period. The reorganisation of institutional structures, therefore, positively affected the ability of users to influence the policy formulation process. At the same time, however, the TOs remained influential; they were still important stakeholders, as they continued to own and operate the public telephone networks. Thus, the Commission clearly could not simply ignore their views. Nevertheless, consistent with historical institutionalism, which asserts that the relative power distribution of social groups is affected by institutional structures (Hall 1986, p. 265), the reorganisation of the institutional setting negated the dominant position that the TOs (former PTTs) had maintained over policy formulation.

In line with the argument put forth by historical institutionalists (March and Olsen 1984, p. 738) that institutions are 'political actors in their own right', the Commission's disproportionate responsiveness to users can be interpreted as a bias in favour of user interests, explained by the Commission's desire to increase its own power, to further the integration process, and to liberalise telecommunications. As Sabatier (1987, p. 662) points out, 'many brokers will have some policy bent'. Nevertheless, the Commission still sought to serve as an effective policy broker and develop reasonable compromise positions between the competing advocacy coalitions. For example, the Commission put forth a revised version of the contentious interconnection provision, which balanced the views expressed by the competing advocacy coalitions. The Commission also set aside its own preference for a mandatory timetable for the provision of advanced services and, instead, put forth a version that accommodated the expressed interests of the TOs.[40]

An analysis of the public consultation period demonstrates that the institutional transformation away from the previously closed and restricted policy community has affected the policy-making process and led to increased openness. Institutional settings, therefore, have an independent and influential effect on policy making. This is not, however, the end of the policy formulation stage. Following public consultation, the Commission discusses its revised proposal with the Member States in the Communications Committee.

4.3(b)   The Communications Committee and National Preferences

The shift from the PTT-dominated SOGT and GAP committees to the NRA-dominated Communications Committee (formerly ONP Committee) has provided for a more liberal EC-level policy network. Because of the institutional modifications mandated by the 1990 Framework and Services Directives, as discussed above, the national delegations of the Member States are headed by the NRAs in these meetings. Also, a Commission official, rather than a PTT representative, chairs these meetings. As a result of the institutional stickiness described in chapter two, however, many of the NRAs maintained close links with their national TOs. National preferences led the newly established French and German NRAs to continue favouring the interests of their national TOs more closely than that of the British NRA, OFTEL. Both France Télécom and Deutsche Telekom remained government owned and continued to contribute to the federal budgets. In contrast to the UK, the federal ministries held on to the main regulatory functions and maintained significant influence over operational decisions.

The purpose of this section is twofold. First, it examines how national preferences, particularly those of the UK, French and German governments, play out in Communications Committee (ONP committee) meetings. Secondly, it continues to assess the extent to which this new institutional setting is less supportive of national TO interests and, conversely, more concerned with user interests, than the GAP meetings had been.

An examination of four key issues[41] raised in the Committee meeting demonstrates that the NRAs generally supported the proposal and were not as strongly opposed to some of the provisions as the TOs had been during the public consultation period.[42] On the first issue examined, which concerned the mandatory provision of advanced services, the policy output from the new institutional setting was in line with the position taken by the more liberal Member States, the United Kingdom and the Netherlands. Since the Commission has the autonomous authority to adopt or reject, in whole or in part, Member State comments at the policy formulation stage, the Commission decided not to accommodate the concerns of the French and German delegations on this issue. Although these two Member States had argued that some of the provisions would be too difficult and too costly to provide, the Commission did not change the substance of the article or the advanced services to which it applies.

Similarly, on the issue of tariff principles, the Commission did not modify the article significantly despite the expressed concerns of the UK, France, Italy and Denmark. However, on the contentious interconnection issue, the Commission expanded the article in line with the comments of the French and German delegations. Since these two Member States had not yet established a comprehensive national framework for interconnection, they sought a detailed text at the EC level. Finally, on the issue of comitology, the Commission, supported only by the UK, reduced the authority of the Member States to oversee the implementation of the proposal. Under the committee procedure put forth in the revised proposal (a type I advisory committee), the Commission maintains autonomous control over implementation, while the Member States can only put forth non-binding advice. This change demonstrates both the Commission's autonomous ability to draft legislative texts in the policy formulation stage and its own preference for increased institutional authority.

Because of its institutional mandate under the EC Treaty to initiate policy, the Commission is thus able to overlook comments and concerns expressed by the Member States and put forth its preferred policy line on key issues to the Council and the EP. As a result of the institutional transformation from the GAP Committee to the Communications Committee (and the national-level separation of the PTTs into separate NRAs and TOs), the former PTTs lost further control over the direction of European telecommunications policies. The institutional transformation of the policy formulation process has thus affected policy outcomes. In line with historical institutionalism, therefore, particular institutional configurations privilege certain actors over others and have an independent and influential effect on policy making.

Following the discussions in the Communications Committee (ONP Committee), the Commission amends the proposal and forwards it to both the European Parliament and Council for the inter-institutional bargaining stage of the policy-making process, which 'has traditionally supported the TO' (Austin 1994). Since the early 1990s, however, procedural modifications to the inter-institutional bargaining stage have increased the authority of the European Parliament. The next chapter examines these procedural modifications and, more specifically, the European Parliament's increased ability to promote user interests.

4.4 CONCLUSION

The policy networks approach is a valuable tool for analysis of the sub-systemic level. From 1987 to 1990, national PTT representatives formulated EC-level harmonisation legislation within a policy community; a balance of power existed with a very limited number of participants with similar resources and a shared economic interest in maintaining the *status quo*. Change in the institutional setting in the early 1990s nevertheless promoted change in the policy network. The formation of the Communications Committee, the ETP, and the three-month public consultation period, along with the introduction of other consultative mechanisms and forums, gave further impetus to the development of a European-wide structure of interest groups; these groups transferred resources to the European level as power shifted. As a result, a large, transnational, loosely organised network of governmental and non-governmental actors, and a more pluralistic policy-making environment, developed. The previously insulated policy community of national PTTs was opened while a new issue network, centred on the Commission, was institutionalised at the European level.

By the early 1990s, after several decades of policy entrepreneurship and battling with the national PTTs (see previous chapter), the Commission finally gained a central coordinating role in the telecommunications sector. Concurrently, the Commission established an elaborate institutional structure to facilitate EC-level interest representation. Due to increasing institutionalisation of the EU's sub-systemic level (Mazey and Richardson 2001), institutional analysis has become increasingly important for explaining and analysing policy formulation. Pluralistic policy-making structures and procedures were established which permitted users, as well as Member States and their national telecommunications operators (formerly their national PTTs), to help shape draft legislation. The reorganisation of institutional structures positively affected the ability of users to influence the policy process; institutions, therefore, have an independent effect on policy making.

As the Commission's policy-making role has expanded in the telecommunications sector, the Commission has become increasingly reliant on independent experts and competing interests for information and advice during the early stages of policy formulation. Other actor-based approaches, such as the epistemic communities and advocacy coalitions approaches, have also become more relevant for analysis of the EC's sub-systemic level. Although advocacy coalitions have traditionally described groups of actors with shared deep core beliefs and values (Mazey 2001), as mentioned previously, there is a strong correlation between beliefs and interests. Sabatier (1987 pp. 663–664) concludes that beliefs are 'more inclusive' than interests and that belief-system models, such as the advocacy coalition approach, 'can thus incorporate self/organizational interests'. Further support for this chapter's argument, that users formed an advocacy coalition rather than merely an *ad hoc* coalition of shared interests, is that they have continued to push for a variety of policies since the late 1980s. This suggests that this group can be characterised by its longevity, which is necessary to obtain the status of an 'advocacy coalition' (Sabatier suggests 10 years). Thus, whether united by a deep core ideology of economic liberalism or merely shared interests, users – joined by some Commission officials and independent experts – formed a stable advocacy coalition, which was able to influence ONP policy formulation throughout the 1990s. At the same time, the Commission acted as a 'broker' between competing advocacy coalitions (albeit with a significant policy bent in favour of user interests) in an effort to formulate mutually acceptable policies.

NOTES

1. In this way, the boundary between the sub-systemic and systemic levels of analysis is clearly defined. In contrast, chapter one found, and Peterson himself admits, that his description of the sub-systemic level is messy and difficult to use theoretically (Peterson, 1995a).
2. Interview, telephone, October 1999. As explained in chapter three, the CEPT is the intergovernmental organisation that the national PTTs established in 1959 in an effort to preserve their autonomy and limit EU encroachment on their work.
3. Interview, UK user group representative, telephone, October 1999.
4. As explained in chapter two, the UK pursued radical reform in the early 1980s and Germany introduced some reform in 1989.
5. Interview, DG Information Society (DG XIII) official, Brussels, May 1999.
6. This is the same definition of users put forth in EU telecommunications legislation (see 1995 ONP Voice Telephony Directive, OJ L 321). Even though private service providers are suppliers to the end users or consumers, they are generally considered users of the public network and, therefore, sided with end users in favouring increased liberalisation in opposition to the national PTTs. For the purposes of this chapter, users will also include those organisations, such as consumer, user and business groups, that represent users in the political arena.
7. Interview, DG Information Society (DG XIII) official, Brussels, May 1999.
8. Interview, UK user group representative, telephone, October 1999. See also Analysys Consultants (1989), p. 61.
9. INTUG (1989) 'Telecommunications Initiatives related to the Single Market in Europe: Background Status and Action Steps'
10. Interview, DG Information Society (DG XIII) official, Brussels, May 1999.
11. Interview, UK user group representative, telephone, October 1999.
12. CEC (1990) 'Commission directive of 28 June 1990 on competition in the markets for telecommunications services (90/388/EEC; OJ L192/10, 24.07.90).
13. Council (1990) Council Directive of 28 June 1990 on the establishment of the internal market for telecommunications services through the implementation of open network provision (90/387/EEC, OJ L192/1, 24.07.90).
14. In 1990, when established, the committee was named the ONP Committee. In 2002, the Communications Committee replaced the ONP Committee and another committee of NRA representatives, the Licensing Committee.

15. The extent to which this institutional modification reduced the ability of the former PTTs (that is, national telecommunication operators (TOs)) to control the policy-making process will be discussed later in this chapter. For an empirical account of the features that distinguish different types of EU committees, see Egeberg *et al.* (2003).
16. Interview, UK user group representative, telephone, October 1999.
17. Interview, UK user group representative, telephone, October 1999.
18. The ETP was initially named the ONP Consultation and Coordination Platform (ONP-CCP). In 1998, the ONP-CCP merged with the European Interconnect Forum (EIF) and was renamed the European Telecommunications Platform.
19. ONPCOM91-40 Memorandum of Understanding and Terms of Reference of the ONP Consultation and Coordination Platform (OCCP).
20. Interview, INTUG-EUROPE representative, Brussels, June 1999.
21. Interview, DG Information Society (DG XIII) official, Brussels, May 1999.
22. Interview, INTUG-EUROPE representative, Brussels, June 1999.
23. Interview, INTUG-EUROPE representative, Brussels, June 1999.
24. Interview, INTUG-EUROPE representative, Brussels, June 1999.
25. Interview, Commission official, Brussels, May 1999.
26. Interview, Brussels, April 1999
27. Interview, INTUG-EUROPE representative, Brussels, June 1999; Interview, UK user group representative, telephone, October 1999.
28. Interview, DG Information Society (DG XIII) official, Brussels, May 1999; Interview, Commission official, Brussels, March 1999.
29. Sutherland, P. (1992) *The Internal Market After 1992: Meeting the Challenge*, Report to the European Commission by the High Level Group on the Operation of the Internal Market (Brussels: Commission of the European Communities), p. 30 quoted in Greenwood (1997), p. 37.
30. CEC (1992) *Increased Transparency in the Work of the Commission*. SEC (92) 2274 (Brussels: CEC); CEC (1992) *An Open and Structured Dialogue between the Commission and Interest Groups*. SEC (92) 2272 (Brussels: CEC); see Greenwood (1997), p. 37.
31. See Commission (1991f); Analysys Consultants (1992)
32. 'Intelligent Voice Services', a report for the Commission prepared by Margrit Sessions of Logica (1990); 'Tariff Principles for PSTN in the Context of ONP'; 'ONP Tariff Guidelines versus Tariff Structures for ISDN'; 'PSTN Productline', all prepared by Wim van Dijk of the European Telecommunications Consultancy Organization (1990).
33. 'Study of the application of the ONP concept to voice telephony services', a report prepared under contract for the Commission by National Economic Research Associates, London (March 1991).
34. Interview, INTUG-EUROPE representative, Brussels, June 1999; Interview, UK user group representative, telephone, October 1999.
35. Interview, DG Information Society (DG XIII) official, Brussels, March 1999.
36. This section on the public consultation period is based partly on the section 'Voting Patterns: ONP Voice Telephony Directive' in Austin (1994).
37. CEC (1991c) ONPCOM91-67, 22 November 1991. The transformation of the Analysis report into the November 1991 proposal, as discussed further below, demonstrates that the modifications to the institutional setting discussed above in section 4.2 enabled the Commission to take account more of the concerns of users.
38. See Appendix 1 for a more detailed analysis of these issues.
39. Interview, DG Information Society (DG XIII) official, Brussels, March 1999.
40. See Appendix 1 for a more detailed discussion of this issue.
41. See Appendix 2 for a more detailed analysis of these issues.
42. For example, although all of the TOs that submitted comments on compulsory compensation (section 2.4 of the Analysis report) during the public consultation period expressed serious concern over the provision, the NRAs generally did not oppose, and several clearly supported, this provision in the Committee meetings. See Commission (1991d), pp. 3–4. See Appendix 2 for a more detailed analysis of this issue.

5. Inter-institutional bargaining in European Union telecommunications

Conciliation has profoundly affected the institutional balance: Parliament and the Council have both had to change their habits in order to give effect to this innovation created by the Treaty of Maastricht.[1] *EP activity report on the co-decision procedure, May 1999.*

INTRODUCTION

This chapter examines the systemic level, or the inter-institutional bargaining stage, of the EC's telecommunications policy-making process. It finds that the European Parliament's (EP's) ability to influence outcomes of the telecommunications policy-making process has increased significantly as a result of procedural modifications to the institutional environment during the late 1980s and 1990s. This chapter comprises two substantive sections. The first section examines EP influence under the cooperation procedure (in force for internal market legislation from 1988 to 1993), through a systemic level analysis of the ONP Framework Directive. The second section examines EP influence under the co-decision procedure (in force since 1993) through a systemic level analysis of the ONP Voice Telephony Directive.

This chapter focuses on the systemic level of analysis, in a temporal sense, as the inter-institutional bargaining stage beginning when the Council and EP receive a draft proposal from the Commission. In the systemic level, structural factors dictate that policy proposals work their way through a complicated set of EC rules and procedures which govern the roles played by the Commission, the EP and the Council. Rational choice institutionalism is drawn upon to examine and explain policy development at the EC's systemic level. It is well suited to analysing the constraints faced by a limited number of interest-maximising, strategic actors as they manoeuvre their way through a complicated set of EC treaty rules and procedures. At the systemic level, Commission officials share formal agenda-setting authority with national government representatives and Members of the European Parliament (MEPs). In particular, it is argued that MEPs have gained agenda-setting abilities, especially when there is a lack of consensus in the Council.

Although most systemic level activity is channelled through a complex institutional framework, 'the EU is not yet a state with a stable institutional system' (Armstrong and Bulmer 1998, p. 61), as reflected in the EP's changing powers. As such, policy making takes place in the midst of unsteady and sometimes vague or contentious EC treaty rules. Within this framework, MEPs and Council Members seek to defend or expand their own powers, frequently expending resources in an attempt to influence the design of the EC's institutions. This unstable political environment has led to delays in the adoption of important telecommunications legislation.

5.1 THE COOPERATION PROCEDURE

Accounts of European telecommunications policies prior to the 1990s tended to ignore the 'very weak European Parliament' (Schneider and Werle 1990, p. 83). Nugent (1994, p. 174) noted that the EP 'has generally been regarded as a somewhat ineffective institution'. Likewise, chapter three did not find the EP to be a major player in European telecommunications prior to the late 1980s. Nevertheless, during the late 1980s and early 1990s, successive modifications to the EC policy-making process have transformed the EP into an important institutional actor in the sector. The 1987 Single European Act introduced a policy-making procedure, the cooperation procedure, which is outlined in Article 252 of the EC Treaty (formerly Article 189c).[2] This procedure permits Council decisions to be taken by qualified majority voting (instead of unanimity) and increases formal EP authority through the introduction of a second reading stage (as outlined in Table 5.1).

Table 5.1   Overview of the cooperation procedure

STEP 1	Draft directive received by Council and EP
STEP 2	EP first reading opinion
STEP 3	Council first reading common position
STEP 4	EP second reading opinion
STEP 5	Commission re-examines proposal
STEP 6	Council second reading

Source: EC Treaty, Article 252.

Although Commission officials, MEPs and Council representatives manoeuvre to maximise personal utility, their options are constrained because they are operating within a 'joint-decision system' (Scharpf, 1997) outlined by the EC Treaty. The complicated set of EC Treaty rules governing the inter-institutional bargaining stage 'establish(es) the conditions for bounded rationality, and therefore establish(es) a "political space" within which many interdependent political actors can function' (Peters 1999, p. 44).

The Commission plays a key role as a policy broker throughout the EC's systemic level (Burns 2004). As argued in the previous chapter, the Commission has established itself as a central actor in the coordination of European telecommunications policies. However, it does not act alone. As Westlake (1994, p. 37) explains, the Commission:

> must draft its proposals with an eye to what will 'play' in Parliament, as well as in the Council. It is involved in the Council deliberations leading to the Common Position ... It must give Parliament its opinion on the Common Position. It must assist Parliament in its deliberations on the Common Position.

In focusing on the Commission, the EP and the Council, this chapter does not consider two other European institutional actors: the Economic and Social Committee (ESC) and the Committee of the Regions (COR). Although the ESC and COR may have a small degree of influence over EC telecommunications legislation, they are consultative bodies and lack formal policy-making authority.[3] In accordance with Scharpf's rational choice institutionalism (1997, p. 71), therefore, this chapter examines 'primary policy actors that are directly and necessarily participating in the making of policy choices'. The focus on key institutional actors in the latter stages of the policy-making process is consistent with my own empirical work and that of Mazey and Richardson (2001) and Peterson and Bomberg (1999), which find that most lobbying occurs in the early stages of the policy-making process and is focused primarily on the Commission (see chapter four).

5.1(a)   The ONP Framework Directive

First European Parliament reading
A micro-institutional view of the EP suggests that the standing committee in charge of telecommunications policy, the Economic and Monetary Affairs Committee (EMAC), has a significant influence on the voting pattern of MEPs and, as a result, on systemic level telecommunications policy making. After the Commission forwarded the draft ONP Framework Directive (discussed in the previous chapter) to the EP and the Council in January 1989,[4] the main part of the EP's legislative work was undertaken by EMAC. The committee's spokesperson or *rapporteur* for the draft text, Belgian MEP

Fernand Herman, issued a report that advised the MEPs as a whole how they should vote in plenary session.[5] The EP, in May 1989, duly approved 18 EMAC amendments by a large majority (see Table 5.2). The fact the MEPs followed Herman's recommendation tends to support the conclusion of a recent study by Benedetto (2005), which found that 'rapporteurs are the most powerful of parliamentarians in terms of influencing the content of legislative outcomes'. This may be particularly true in telecommunications, where, as discussed below, many MEPs are willing to rely on EMAC's recommendations because they do not have the technically specialised knowledge to understand telecommunications legislation.

Indeed, only a small fraction of MEPs belong to the epistemic community with 'recognised expertise and competence' (Haas 1992, p. 3) over telecommunications issues. This confirms the value in a 'knowledge perspective' (Radaelli 1995, p. 178) or, in other words, an examination of the interaction between knowledge and important policy-making features, such as decision rules, policy problems and policy styles. Most officials interviewed for this book estimate that only eight to ten MEPs fully understand EC telecommunications policies, and that nearly all of these MEPs are found on the EMAC.[6] Other MEPs, therefore, rely heavily upon the opinions of this committee.[7] Because each MEP covers a wide range of issues, they are generally not as specialised as Commission officials or Member State representatives, and therefore must often rely on colleagues for information and advice.

MEP Fernand Herman explained, 'for telecommunications legislation, you have a particular phenomenon, nobody understands [it] ... That's why most of the members of the Parliament trust their colleagues [on the EMAC committee] which are specialised in this (area).'[8] The EMAC committee has a few Socialists, a few Christian Democrats and a few Liberals who are 'aware' and 'experienced' in telecommunications issues.[9] Because of this, the 'EMAC committee is rather well-balanced, in other words [after] we agree [a position] in EMAC, there is not a lot of change to get the whole assembly to accept it'.[10] This gives the telecommunications experts on the EMAC committee considerable agenda-setting authority in the inter-institutional bargaining stage. It also demonstrates the importance of knowledge in the policy-making process; within the EP, 'power and knowledge perform complementary functions' (Radaelli 1995, p. 159). EMAC members are thus important actors in telecommunications policy making; their resource is detailed knowledge of telecommunications issues.

The Commission, which has the authority to alter draft proposals throughout the inter-institutional bargaining stage, put forth a revised proposal in August 1989 incorporating most EP amendments (see Table 5.2).[11] Some of the changes incorporated into the Commission's revised

Table 5.2 Commission and Council first reading reaction to EP amendments May 1989

EP AMENDMENTS (26 May 1989)	COMMISSION REVISED PROPOSAL (10 August 1989)	COUNCIL COMMON POSITION (5 February 1990)
Nos 42 and 43: TO ESTABLISH an advisory committee (Communications Committee) which will consult the TOs, users, consumers, manufacturers and service providers in place of *the SOGT.*	followed EP	similar wording, which is close in spirit to the EP amendment
No 2: TO PURSUE multilateral negotiations aimed at furthering market opening **preferably** in the GATT in place of *in particular* in the GATT.	similar to EP	similar wording, which is close in spirit to EP amendment
No 26: TO DEFINE 'telecommunications organizations' as **public or private bodies … to which a Member State grants special or exclusive rights** in place of *administrations or private operating agencies …*	modified in direction of EP Amendment	followed COMM revised text (and that of EP), but introduced a completely new second paragraph
No 4: TO DELETE the provision enabling the Council and the Commission to supplement bodies considered to be 'Telecommunications organizations.'	abandoned original text and came up with new text	reworded the new Commission text
No 25: TO ENSURE the **comprehensive and effective** protection **and confidentiality** of data.	did not follow EP	did not follow EP
Nos 6, 8, 10, 11 and 12: TO ENSURE a role for the EP, **pursuant to Art. 100a,** in expanding and further defining ONP conditions, basic principles and essential requirements.	followed EP	followed EP

104

No 7: TO ENABLE the Commission, in consultation with an **advisory committee**, to determine service interoperability and data protection as essential requirements	followed spirit, if not wording, of EP text	partly followed EP, but decided upon a regulatory committee
No 37: TO MAINTAIN the solvency of the TOs	did not follow EP text	did not follow EP
No 13: TO SUBMIT a report to the Council **and EP** in 1992 on the need for the **further opening** of the EC telecommunications market in place of the *further harmonisation*	followed spirit of EP amendment to include both further opening and harmonisation	modified Commission text, but still in accordance with spirit of EP amendment
No 31: TO ADOPT ONP conditions for new **ways of using the** network infrastructure … in place of new *types of access to the local network* infrastructure …	did not follow EP	deleted entire clause
No 40: TO DELETE many of the access and supply conditions for which harmonised ONP usage conditions could be developed	did not follow EP	did not follow EP
Nos 41 and 15: TO AMEND the reference framework for the definition of harmonized tariff principles	did not follow EP	did not follow EP, and modified COMM text

Notes: *Italics* indicate the initial Commission wording; **Bold** indicates actual textual modification proposed by EP.

Sources: European Parliament (1989); Commission of the European Communities (1988b); the EMAC draft recommendation on common position, PE 139.188, 29.3.90; and Council (1990).

proposal are part of the institutional adjustment to the policy formulation stage that took place in the early 1990s (see chapter four). In particular, the Commission took aboard EP amendments that established the more liberal Communications Committee (formerly the ONP committee) in place of the PTT-dominated SOGT and GAP committees and also put user representatives, manufacturers and service providers on equivalent footing with the TOs.

Crucially, these amendments helped develop the more pluralistic policy-making environment examined in the previous chapter. In line with the self-interested nature of the MEPs, other amendments enhanced the EP's policy-making role. Commission officials, however, rejected other EP amendments because they felt the Framework Directive was not an appropriate place to address issues such as data protection and TO solvency.[12] Nevertheless, the fact that the Commission accepted most EP amendments, some of which significantly changed the proposal (for example, enhancing user input into policy formulation), suggests that the EP had at least some influence on the proposal during its first reading.

First Council reading
Since Member State representatives are rational actors seeking to defend national interests at the EC level, the outcome of the Council negotiations is best explained through an examination of national policy preferences. The main disagreement had been over whether to include a provision for the development of ONP conditions for data services. This disagreement resulted from differences over national perspectives on the list of services that should be entrusted specifically to the national TO (public telecommunications services).

The UK had liberalised in the early 1980s (see chapter two) and thus maintained a narrow definition of public telecommunications services. Germany had already established *Postreform I* earlier in 1989 (see chapter two), which liberalised value-added telecommunications services, including data transmission services. The UK and Germany, therefore, with the support of the Netherlands and Denmark (both of which had liberalised data services), sought to restrict the scope of public telecommunications services to the provision of voice telephony.

France, along with the other seven national delegations, on the other hand, had not introduced significant structural reform, continued to be represented by their PTT Ministers in the Telecommunications Council and had not liberalised data transmission services. Although Hubert Prévot, the senior civil servant who headed France's public consultation, had issued his recommendations for reform in August 1989 (see chapter two), it was not until mid- to late-1990 that a series of laws and decrees for reform were introduced in France. PTT officials continued to argue that their monopolistic provider had successfully established one of the most advanced telecommunications systems in Europe. France thus led the other seven Member States in arguing for a more expansive scope for public telecommunications services and, particularly, the inclusion of data transmission services.

The political sphere under which these Member States negotiate is defined by the EC's joint-decision system. Such systems are 'compulsory negotiation systems that, in the multiparty case, may also be characterized as collective-decision or voting systems operating under either unanimity or consensus rules' (Scharpf 1997, p. 144). In the Council, a majority of decisions are still made by consensus even after the SEA, when qualified majority voting was introduced. What is different, however, is that if the Member States should fail to reach a consensus, the majority can now get its way (Scharpf 1997, p. 144). In the case of the Framework Directive, however, national representatives failed to achieve either a consensus or a qualified majority in favour of the proposal. Under the EC treaty rules for qualified majority voting, the UK, Germany, the Netherlands and Denmark formed a blocking minority. Thus, negotiations on the draft directive had stalled.

Highlighting the key role that the Commission can play in facilitating agreement between the Member States, towards the end of 1989, Commission officials threatened to resort unilaterally to the Commission's competition authority if national representatives were unable to reach an agreement on the Framework Directive (see chapter seven). As a result of this Commission pressure, along with the mounting technological and economic pressures for reform (see chapter two), and hard bargaining between the Member States, a political compromise was reached in the Telecommunications Council in December 1989. The 12 Ministers agreed to a provision that allowed the voluntary application of ONP conditions for the technical interfaces of data transmission services. These harmonised technical interfaces were to be developed by the ETSI, and strongly recommended to the TOs (Council 1990, arts. 4.4(c) and 5.2; Wheeler 1992, p. 85).

The Member States were thus only able to reach agreement on the basis of unenforceable, soft law. This indicates how difficult policy change can be, where the self-interests of national governments determine all possibilities of institutional development (Scharpf 1988, pp. 267–268). Nevertheless, the Commission, in consultation with the Member States, could make the European standards compulsory if implementation of voluntary standards appeared inadequate (Council 1990, article 5.3). In line with the two-pronged strategy outlined in the 1987 Green Paper (see chapter three), the Council also agreed to link liberalisation of the services market, through the entering into force of the Article 86(3) (formerly Article 90) Services Directive (see chapter seven), to the final adoption of the Framework Directive.[13]

Following this agreement, a common position was adopted at the February 1990 Telecommunications Council. Taking advantage of the vagueness of the EC Treaty Article 252 (formerly Article 189c) requirement that the Council inform the EP 'fully of the reasons which led the Council to adopt its common position', Council members issued a brief statement that there were 'particularly long and difficult discussions within the Council' and that in negotiating the final compromise set out in the common position, 'the Council had to overcome deeply diverging views on the potential scope of ONP'.[14] The only explanation given was that these views 'were linked to the concept upheld by each Member State regarding the role of state intervention in this area'.[15]

In this case, Ministers clearly felt that it was in their collective self-interest to withhold the details of the difficult political compromise from the EP and the public. One MEP argued that the lack of information disadvantaged the EP in reacting to the common position and weakened its position vis-à-vis the Council.[16] Nevertheless, where EC Treaty rules are vaguely defined, strategic rational actors will often interpret these rules consistently with their perceived best interest and current institutional structures and norms, and in a way that maximises their own authority. Since 'the rules are often vague, contentious or "shiftable"' (Peterson and Bomberg 1999, p. 254), procedural disputes in the EC's systemic level are not infrequent. As is argued further below, this can lead to less efficient, 'sub-optimal' (Scharpf, 1988) policy making, since 'the "reluctant Europeans" among member governments have been much more willing to accept disagreeable compromises on substantive policy than to weaken their own institutional control over the substance of future decisions' (Scharpf 1988, p. 268).

The Council accepted 12 of the 18 amendments put forth by the European Parliament (see Table 5.2). These amendments changed the proposal in significant ways: EP prominence in telecommunications policy making was increased through a formal role in developing ONP conditions; TO prominence was reduced through the elimination of the SOGT committee; and the definition of TO was broadened to include all operators granted special or exclusive rights. This suggests that the EP, supported by the Commission, had some institutional influence in its first reading.

There are, however, three further steps to the cooperation procedure. These final steps are governed by a more complex set of EC Treaty rules and procedures (as summarised in Table 5.3) than the first three steps examined above. For example, in contrast to the first readings, the EP and Council are required to respect three-month time limits when re-examining the proposal.[17] It is argued below that these detailed rules and procedures have a significant effect on the strategies and aims of key actors and a significant influence on policy outcomes. As Hall and Taylor (1996, p. 945) point out, 'one of the great contributions of rational choice institutionalism has been to emphasize the role of strategic interaction in the determination of political outcomes ... Institutions structure such interactions, by affecting the range and sequence of alternatives on the choice-agenda.'

Table 5.3   Steps four through six of the cooperation procedure

STEP 4 EP second reading opinion	EP **approves** common position or has not taken decision within 3 months	EP **rejects** common position by an absolute majority (260 MEP) within 3 months	EP **proposes amendments** to the common position by an absolute majority (260 MEP) within 3 months
STEP 5 COMM re-examines proposal	↓	Commission re-examines proposal (one month)	Commission adopts re-examined proposal **with** EP amendments -or- Commission adopts a re-examined proposal **without** the (or certain) EP amendments
STEP 6 Council second reading	Council **adopts** the decision	Council **adopts** common position by unanimity within 3 months -or- Council does not take decision within 3 months (proposal is **not adopted**)	Council **approves** re-examined proposal by a qualified majority within 3 months -or- Council **modifies** re-examined proposal unanimously within 3 months -or- Council does not take decision within 3 months (proposal is **not adopted**)

Source: Adapted from Figure 1 in Tsebelis (1994), p. 130.

Second European Parliament reading, Commission approval and Council adoption

In May 1990, following an EMAC recommendation, the EP duly approved three amendments in its second reading.[18] Because of this, the Commission was required to re-examine, possibly amend and then forward the proposal to the Council again within a month. At this point, the Commission's role as a policy broker is particularly important, because Commission officials have to juggle the concerns of MEPs and Council members in order to reach an agreement. As Nugent (1994, p. 325) states, the Commission must use:

> its judgment as to whether ... it should amend its text so as to get proposals through in a reasonably acceptable form. At the second reading stage, in particular, a delicate balance may have to be struck: between, on the one hand, being sufficiently sympathetic to EP amendments so as not to upset MEPs too much, and, on the other hand, being aware that a revised text might break up a majority attained in the Council at first reading.

In this case, Commission officials were extremely cautious and sought to preserve the Council majority obtained in the first reading. As such, the Commission did not accept any EP amendments. The influential effect that the prevailing rules and procedures (presented in Table 5.3) had on the strategy of Commission officials is demonstrated by an examination of the reasoning behind the Commission's rejection of the EP amendments.

A Commission official noted that the 'main argument was based on the fact that the text of the common position was the final result of a long period of discussion in the Council'.[19] Commission officials were concerned that, if the Commission had accepted any EP amendments, the Council would have needed a unanimous vote to overturn them. The formal rules thus restricted the options available to Commission officials and, at the same time, privileged the Council's role and power over the EP's role and power. 'A unanimous decision would have required a lengthy discussion in the Council working group ... In the end, this very well could have exceeded the time limit of three months, including the summer break, and endangered the whole directive.'[20] It is likely that several Member States, including the UK, Denmark and the Netherlands, would have forcefully opposed the EP amendment aimed at ensuring TO solvency. Furthermore, Commission officials were concerned that, because the Council had agreed that the adoption of the ONP Framework Directive was linked to the automatic entering into force of the Services Directive, several Member States, including Belgium, Italy and Greece, might have chosen not to adopt the ONP Directive in order to block the Article 86(3) Services Directive.[21]

The Commission's decision to reject the EP amendments at the second reading stage can thus be explained, at least partly, by the limits that the underlying institutional arrangement posed to the alternative courses of action that Commission officials considered (see Hall 1986, p. 265). Scharpf (1997, p. 42) explains that rules which permit some actions and proscribe others, 'define repertoires of more or less acceptable courses of action that will leave considerable scope for the strategic and tactical choices of purposeful actors'. Thus, after having thoroughly examined the possible courses of action, Commission officials returned the original common position back to the Council. Although agreeing in principle with the objectives of the EP's amendments, the Commission's opinion stated that the idea behind the amendments had been adequately incorporated into the common position (Commission 1990b). The Telecommunications Council subsequently adopted the common position as a legislative act in June 1990.

Since the Commission did not include EP amendments, at least partly to secure passage of the ONP Framework Directive, it played a key institutional role as a policy broker between the EP and the Council. Under the cooperation procedure, Commission officials knew that the EP could not reject the proposal in retaliation for the Commission's refusal to incorporate EP amendments. However, this would have been a real danger under the institutional environment established under the co-decision procedure (as discussed below). Therefore, although a majority of EP amendments (12 out of 18) were incorporated into the ONP Framework Directive during its *first* reading, the three amendments that the EP put forth in its *second* reading were not accepted. Earnshaw and Judge (1996, p. 124) note that:

> under co-operation, Parliament could be seen metaphorically to be jostling the two other institutions – whispering respectively in their ear, and nudging them considerable distances down the path – before being effectively pushed off onto the verge of decision-making after Council adopts its common position. Parliament's attempts to clamber back onto the path thereafter, through amendment and the occasional formal rejection of the common position, met with only limited success.

Thus, the 'outsider' (Earnshaw and Judge 1996, p. 109) EP was unable to get a total of seven amendments it had put forth incorporated into the directive.[22] The Council's refusal to give the EP adequate reasoning behind its common position also demonstrates a lack of EP institutional authority. Although the cooperation procedure enabled the EP to influence the ONP Framework Directive in significant ways, the procedure still enabled the Council and, thus, mostly the state-owned PTTs, to play the dominant role during the inter-institutional bargaining stage. The institutional balance of power, therefore, remained tilted in favour of Member States.

5.2   THE CO-DECISION PROCEDURE

Following ratification of the Maastricht Treaty in 1993, the EP gained further legislative power under the co-decision procedure, as outlined in EC Treaty Article 251 (formerly Article 189b). Whereas the legislative process for the ONP Framework Directive left the formal adoption up to the Council, subsequent telecommunications legislation has been adopted under the co-decision procedure in which the Council shares formal legislative authority with the EP. Craig and De Búrca (1995, p. 125) note that:

> the TEU has introduced yet another complex procedure which bolsters still further the powers of the European Parliament. This new procedure has been referred to as 'co-decision', both because it is designed to prevent a measure being adopted without the approval of the Council and the European Parliament, and because the procedure within the Article places emphasis on the reaching of a jointly approved text.

The 1997 Amsterdam Treaty simplified and expanded the use of the co-decision procedure in several ways (Maurer 2003). It is nevertheless still fairly complex and has led to disputes over institutional authority.

The first two readings are, for the most part, the same as steps one through six of the cooperation procedure examined previously (see Table 5.1).[23] In the event that the EP votes (in step 4) to reject the Council common position or the Council refuses to accommodate (in step 6) all EP second reading amendments, the co-decision procedure introduces a conciliation committee (steps 7 and 8). The conciliation committee has established a new institutional setting, which has significantly enhanced the ability of the EP to influence the telecommunications policy-making process after the Council adopts its common position (after step 3).

5.2(a)   The ONP Voice Telephony Directive

Although also submitted initially to the EP and Council under the cooperation procedure, following the entry into force of the Maastricht Treaty on 1 November 1993, the draft ONP Voice Telephony Directive (see previous chapter) was formally re-submitted under the co-decision procedure.

First European Parliament reading

Following an EMAC recommendation, the EP approved the voice telephony proposal in March 1993 with 37 amendments, mostly designed to safeguard the interests of service providers and consumers (see Table 5.4).[24] In line with the institutionalist view of the EP as a self-interested authority maximiser (Pierson 1996, pp. 132–135), *rapporteur* MEP Mel Read (UK, PES) noted that the additions would enhance Parliament's role in consultations with the Council (*Reuter European Community Report*, 10 March 1993, citing EP Press Release A3–64/93-Read). The Commission then put forth a revised proposal in May 1993 that incorporated most

EP amendments (see Table 5.4). The fact that the Commission once again accepted a majority of the EP amendments suggests that the EP influenced the proposal significantly during its first reading (that is, the institutional rules privilege the EP at this stage).

First Council reading
As discussed in chapter two, the Right returned to government in France after the March 1993 elections. Since this new government favoured far-reaching national reform, French representatives were likewise more supportive of EC-level reform efforts than the previous government had been; there was now a closer fit between national and EU policy objectives (see chapter two; see also Thatcher 1999a, pp. 157–162). Nevertheless, at the Council meeting of May 1993, French officials sought to modify the article on tariff principles. After having failed to bring about the desired changes to this article in the policy formulation stage (see Appendix 2), French officials were determined to try again in another institutional context where they possessed greater power. In particular, French officials sought to provide Member States with the option of effecting tariff equalisation under general policy objectives such as universal service or regional development (*Agence Europe*, 8 May 1993). For example, Member States might levy tariffs higher than necessary to cover land-use planning costs. Thus, even with the changed government, the strongly embedded commitment to voice telecommunications services as a *service publique* continued to influence French policy objectives at the EC level.

With regard to the institutional arrangements for the implementation of the directive, national representatives sought to create an institutional design most favourable to their interests. The French delegation, with support from Spain, Belgium, Luxembourg and Portugal, sought to strengthen national coordination over the Commission and argued that decisions over convergence and technical adaptation should not be left to comitology committee procedures.[25] These five Member States argued that decisions should instead be taken under the Council's normal legislative procedures because of the heavy implications in terms of investment for the companies concerned. The French delegation commented, however, that if it obtained satisfaction on tariff equalisation, it might remove its opposition to the committee procedure (*Agence Europe*, 8 May 1993).

The UK, as was the case in the Communications committee (ONP committee) (see Appendix 2), continued to be the only Member State to favour a type IIIa regulatory committee, which requires a difficult-to-obtain qualified majority for Member States to reject Commission measures (*European Report*, 18 November 1992). Because the UK had liberalised previously, the British government felt that the Commission was on its side and therefore:

> a less powerful [committee], as far as the Member States are concerned, is okay. The Commission needed to do more things and the Member States cannot frustrate the Commission [in a type IIIa committee]. The Commission is able to play its liberalising role, which is what we wanted it to do. So in this particular situation, a weak committee for us was fine.[26]

Thus, in this instance, UK preferences and policy style were consistent with the Commission aims. Meanwhile, the German delegation no longer argued against the committee procedure as it had done 18 months earlier in the policy formulation stage (see Appendix 2). Mounting economic pressures stemming from German reunification, combined with continued domestic resistance to reform of the Basic Law (see chapter two), effectively reduced the opposition of German officials to Commission authority to pursue reform.

The Member States reached agreement on a common position in May 1993 (Common Position No. 247/93). In line with the preference of the UK delegation, the common position provided for the type IIIa regulatory committee to take binding measures on the most important aspects of implementation if convergence was insufficient (*Agence Europe*, 8 May 1993). Spain abstained and Portugal voted against the common position because they both continued to oppose the type IIIa regulatory committee (Council of Ministers Press Release, 10 May 1993). The French delegation was 'bought off'; it obtained satisfaction over tariff equalisation and therefore voted in favour of the common position (Article 11 and recital 22; *European Report* 12 May 1993; *Agence Europe*, 12 May 1993). It is this type of logrolling ability that France lacked during the policy formulation stage, which demonstrates that a different institutional context led to different winners and losers.

More than half of the EP first reading amendments (20 of the 37) were incorporated into the Council common position (see Table 5.4). The fact that a large number of amendments were accepted into the Council common position once again demonstrates a significant legislative influence during the EP's first reading.

On the other hand, illustrating the importance of early lobbying before Member States took their position, at least 13 amendments concerning the rights of users and the transparency of implementation procedures were not accepted.[27] Although, under the cooperation procedure examined in the first part of this chapter, the EP was not able to get any of its second reading amendments incorporated into the ONP Framework Directive, the following section seeks to determine whether the new institutional context provided for under the co-decision procedure (steps 7 and 8) enhanced the EP's authority to amend the Council's common position.

Second EP reading

Following the position of EMAC, the EP introduced 14 amendments during its October 1993 second reading.[28] Twelve of these amendments reintroduced (either wholly or partly) first reading amendments.[29] In addition, two new amendments were put forth. One amendment aimed to reinsert a provision that was in the original Commission proposal but had been deleted by the Council, requiring TO accounts to be audited annually. The other amendment aimed to eliminate reference to the type IIIa regulatory committee for the follow-up to the directive on the grounds that both the Parliament and Council should be consulted on significant adaptations.

Rapporteur Mel Read noted that the Council's common position made 'comitology ... a more serious issue' because the regulatory committee 'has been traditionally regarded as unacceptable by the parliament'.[30] Following ratification of the Maastricht Treaty, MEPs felt that they should play a greater role in the implementation of co-decision legislation. MEP Biagio De Giovanni (Italy, PES) of the Institutional Affairs Committee explained the wider significance of the role changes:

> the Council used to have a power of delegation to the Commission for the execution of acts it had adopted, but it is quite clear that in the case of acts of co-decision that exclusive power of delegation lapses because the Council no longer has sole responsibility for the act and it is therefore also clear that this power belongs to the Council and the European Parliament jointly.[31]

Once again, in line with the view that MEPs are egoistic and 'concerned not only with influencing policy substance but also with ensuring that the EP's influence is as strong as possible in inter-institutional relations', (Armstrong and Bulmer 1998, p. 79) the MEPs expressed their willingness to veto the legislation if the Council failed to agree to the amendments. Read directly challenged the common position by warning that it was subject to the Union's co-decision procedure.[32] In threatening to resort to their new institutional authority, the MEPs became more aggressive and adjusted their strategy in response to a change in the institutional setting. In demanding a role in the implementation committee, MEPs adjusted their goals and sought to influence the EC's institutional design in a way that would maximise their authority. Therefore, the institutional environment helped to shape the actions, strategies and goals of the MEPs.

Table 5.4   Commission and Council first reading reaction to EP amendments March 1993

EP AMENDMENTS (March 1993)	COMMISSION REVISED PROPOSAL (May 1993)	COUNCIL COMMON POSITION (July 1993)
No 29 TO REQUIRE NRAs to submit details to Commission of cases where access to network has been denied or restricted	followed EP	deleted entire clause
Nos 1 and 31: TO PROTECT users from unreasonable price increases; and expensive procedures for settling disputes	followed EP	modified text, but still in direction of EP amendments
No 2: TO PROTECT TOs from unreasonable access requests and to balance TO concerns with user rights	followed EP	followed EP
No 3: TO ENCOURAGE implementation of fully distributed costing **within a reasonable time limit**	followed EP	followed EP
Nos 4 and 30: TO ENSURE establishment of effective NRAs	did not follow EP	did not follow EP
Nos 5, 6, 19, 25, 28, 34, 35 and 36: TO INCREASE openness, transparency and the role of the EP in procedures for technical adjustment; the convergence of ONP conditions; requesting ETSI standards; the promotion of EC-wide numbering; examining implementation and proposing further measures to implement the directive's aims; and openness and transparency in committee procedures.	did not follow EP	did not follow EP
Nos 7, 11, 32 and 33: TO CLARIFY the definition of the 'public telephone network'; the obligations placed upon the Member States and the TOs in ensuring the provision of voice services; the procedure for convening a dispute resolution working group	followed spirit, if not wording of EP text	modified text, but still in direction of EP amendments
Nos 8, 9, and 10: TO INCLUDE a definition of 'telecommunications organization'; 'special or exclusive right'; and 'small or medium-sized' telecommunications organisations	did not follow EP	did not follow EP

Provision		
Nos 12 and 27: TO FACILITATE the ability of users to obtain information on access conditions and standards	followed spirit, if not wording of EP	Similar wording to Commission text
Nos 13 and 22: TO ENSURE **annual** publication of TO performance and NRA review of service targets **at least every three years**; and TO implementation of approved cost accounting system **within one year**	followed EP	followed EP
No 14: TO ENSURE **affected** user consultation prior to a change in service offerings	followed EP	followed EP
No 15: TO REQUIRE compensation to users for failure to meet contracted service quality levels	followed EP	did not follow EP
Nos 16 and 17: TO REQUIRE TOs to gain NRA approval to vary published tariff and supply conditions; TOs only need to respond to **reasonable** requests and must give a prompt and justified explanation to users when special access is refused	followed EP	partly followed EP, but not requirement for denial-of-access explanation
Nos 18 and 20: TO PREVENT TOs from favouring a particular service provider	followed EP	followed EP
No 38: TO ESTABLISH tariff unbundling so that users do not pay for features which are not **required for the service demanded** in place of *requested*	followed EP	followed EP
Nos 21 and 23: TO PROTECT small and medium-sized TOs in negotiating tariffs for network access to larger TOs; users of small or medium-sized TOs through discount schemes	did not follow EP	did not follow EP
No 24: TO ENSURE itemised billing is **made** available to users in place of available *on request*	similar wording to EP amendment	partly followed EP, but availability is *on request*
No 26: TO PROTECT the less-developed TOs in noting the technical limitations in interrupting particular services for non-payment of bills	followed EP	followed EP

Notes: *Italics* indicate the initial Commission wording; **Bold** indicates actual textual modification proposed by EP.

Source: Commission of the European Communities (1992b); European Parliament (1993); COM (93) 182 final – SYN 437, OJ C 147/12, 27.5.93; Common Position No 247/93 adopted (formally) by Council on 13 July 1993.

Commission and Council re-examine proposal

Telecommunications Commissioner Martin Bangemann explained that the Commission was unable to accept several amendments because they were adequately incorporated into the ONP Framework Directive.[33] In addition, because France, with the support of Spain, Portugal, Belgium, the Netherlands, Ireland and Luxembourg, found the three EP amendments relating to comitology unacceptable, the Commission did not adopt them.[34]

In March 1994, the Commission accepted four of the EP's fourteen second-reading amendments.[35] Nevertheless, because ten of the Parliament's amendments were not accepted, the Council President, with backing from the EP president, convened a conciliation committee (*Agence Europe*, 16 March 1994). The EC Treaty rules and procedures that govern the convening of the conciliation committee and the subsequent voting of the EP and Council are outlined briefly in Table 5.5. The following section contends that these rules and procedures help to shape the politics and strategies of policy makers and play a large part in determining the policy outcome.

Table 5.5   Steps seven and eight of the co-decision procedure

STEP 7   Council and EP presidents convene a conciliation committee. The conciliation committee consists of an equal number of representatives from the Council and Parliament. Commission representatives attempt to reconcile disagreements between the two delegations. The conciliation committee has six weeks to approve a joint text.
STEP 8   Conciliation committee **approves** joint text. The EP and the Council each have six weeks to approve the text (by absolute and qualified majority respectively). If either institution fails to gain approval, the proposed act is *not* adopted. -or- Conciliation committee **does not approve** joint text. The proposed act is *not* adopted.

Source: EC Treaty, Article 251 (formerly Article 189b).

Informal conciliation

The Greek Presidency of the Council met Parliament's *rapporteurs* to lay the groundwork for discussions in the conciliation committee. MEPs and national representatives place a great deal of importance on these pre-conciliation committee, informal gatherings. These meetings have emerged as crucial arenas for informal bargaining. Informal conciliations are held about a week before formal conciliation committees meet and usually are relatively small, intimate gatherings designed to agree agendas for formal committee meetings ... [however] an EP official suggested that informal conciliations 'can be more important than formal conciliations'. They allow people to say: 'Here's our informal agenda. Do you have one?' (Peterson 1995a, p. 87)

The informal nature of these gatherings, and the small number of participants, facilitates bargaining between the EP and the Council. Demonstrating that informal conciliations have become an important *ad hoc* institution for reducing areas of conflict, the EP and Council delegations reached agreement on most of the amendments. Following the discussion, however, the Greek Presidency indicated that the three amendments on comitological issues and the amendment on compulsory compensation (no. 15) were still problematic (*European Report*, 30 March 1994). Because the two delegations were unable to negotiate a compromise on *all* amendments in this informal institutional setting, the EP and Council would find it more difficult to reach agreement on a joint text in the larger, more cumbersome and formal conciliation committee.

Conciliation committee

In contrast with the informal pre-conciliation meetings discussed above, the large number of participants in the formal conciliation committees complicates negotiations. The Council delegation includes a representative from each of the 12 Member States. The EP president, a vice-president, eight members nominated by the political groups, the *rapporteur*, and the relevant committee chairperson represent the EP delegation. Parliament's standing committees, such as EMAC, are not directly involved in conciliation meetings.[36] MEP vice-president Nicole Fontaine argues that, although the procedure is 'rather cumbersome and complicated', the 'unity' of the EP delegation is generally its 'greatest weapon'. The Council delegation, on the other hand, is less unified and rarely represented by Ministers (*Reuter European Community Report*, 22 February 1995). MEPs contend that the disunity of the Council delegation can make it difficult to reach agreement on a joint text. MEP Fernand Herman explains, 'It's a very difficult dialogue ... very often [the Council President] has to stop the meeting for referring back to Paris, Bonn, to London ... it's really frustrating.'[37]

The EP and the Council delegations held formal conciliation meetings on the voice telephony proposal on 29 March 1994 and 26 April 1994. As in the informal gathering discussed above, compromise was within reach on most EP amendments. The Council delegation was flexible on several amendments, while the EP delegation was also willing to compromise on some of its demands.[38] The two sides, however, continued to disagree on two issues. First, the EP stood firm in its support of compulsory compensation (no. 15), while the UK, with the support of the Netherlands and Belgium, firmly opposed it. Both the UK and the Netherlands had long-established deregulated markets and thus viewed this additional regulation as a step backwards.

Second, highlighting the importance these utility-maximising actors place on institutional design, the EP and Council delegations continued to disagree over committee procedures.[39] The EP delegation argued that recourse to a type IIIa regulatory committee was incompatible with the framework of the co-decision procedure because it would undermine the EP's status. France, with the support of Belgium, the Netherlands, Ireland and Luxembourg wanted a greater degree of national control over the Commission through the type IIIa regulatory committee and thus opposed the EP comitology amendments. Spain and Portugal, on the other hand, opposed both the EP comitology amendments and the type IIIa regulatory committee, in favour of the normal legislative procedures that give the most authority to the Member States (*European Report*, 30 March 1994). Spain and Portugal, as two of the least liberalised Member States, wanted to ensure the greatest possible degree of national control over the implementation of the directive.

Demonstrating the extent to which these self-interested actors seek to influence the EC's institutional design, the entire EP delegation asserted that the dispute over the design of the implementation committee was the only issue preventing the two sides from reaching agreement on a joint text (*European Information Service*, 12 April 1994). Because the EC is still 'a relatively young and experimental system of governance', (Peterson and Bomberg 1999, p. 21) and the EC treaty does not define precisely every aspect of inter-institutional interaction, conflicts over the design of institutions, rules, and procedures are a salient feature of the inter-institutional bargaining stage. As Monar (1994, p. 695) points out, the

> institutions frequently use all political and legal means available to increase their impact on the decision-making process or defend their prerogatives. This interinstitutional dynamic is enhanced by the framework (or lacunae) character of the EC treaties which leaves much room for different interpretations of the political role to be played by the institutions and on how to apply certain procedures.

The Commission can play an important role in the conciliation meetings as a broker between the two delegations. The Commission takes 'all the necessary initiatives with a view to reconciling the positions of the European Parliament and the Council'.[40] The Commission representative attempted to broker a solution to the inter-institutional dispute by proposing separate negotiations on the reform of the 1987 Comitology Decision.[41]

Since MEPs did not have a formal role in implementation committees under the 1987 Decision, the Commission representative agreed with MEPs that the Decision was not suited to the post-Maastricht legislative environment. Nevertheless, the Commission representative recommended that the EP tentatively approve the existing provisions until an agreement on new procedures was reached. In this instance, however, the Commission was an unsuccessful policy broker because the EP delegation maintained that the Commission proposal meant acquiescing to an unfavourable institutional design (that is, the type IIIa regulatory committee). The MEPs therefore rejected the Commission's initiative (*European Information Service*, 12 April 1994).

Council and EP third readings
Subsequent to the 1997 Amsterdam Treaty, failure to reach agreement in the conciliation committee means that the proposed act is not adopted (see Table 5.5). However, because the inter-institutional bargaining over the instant voice telephony proposal took place prior to the Amsterdam Treaty, the Council and EP were each entitled to a third reading of the proposal. In a further attempt to broker a resolution to the inter-institutional dispute during this third reading, Commissioner Bangemann urged the Council to make a 'gesture of goodwill' and accept some of the 14 amendments (*European Insight*, 10 June 1994). The Commission thus reintroduced its modified common position, which included four EP amendments (see above). Demonstrating, once again, that inter-institutional conflicts over procedures are a salient feature of the EC's systemic level, the Commission issued a declaration suggesting that, if the Council refused the four amendments, the common position should be confirmed by a unanimous vote (*European Report*, 28 May 1994).[42]

The Council, on the other hand, pointed out that the EC Treaty Article 251 (formerly Article 189b) stipulates that the 'Council, acting by a qualified majority ... confirms the common position which it agreed before the conciliation procedure was initiated, possibly with amendments proposed by the European Parliament'. The Council argued that the common position was an act of Council and that the Commission could not amend the proposal (Earnshaw and Judge 1993, 1996). In consequence, the 'ministers [were] as one in feeling that the confirmation should be carried out by a qualified majority, not unanimity' (*European Report*, 28 May 1994). Thus, the procedural dispute between the Council and the EP was compounded by a procedural dispute between the Council and Commission.

Although the EC Treaty defines a great deal of the interactions between the institutions during the inter-institutional bargaining stage, these disputes demonstrate that there are inevitably some vague aspects of the EC policy-making framework. March and Olsen (1989, p. 58) contend that, 'although the rules and routines of institutional life are relatively stable, they are incomplete. It is possible to influence the resolution of ambiguity surrounding the rules.' Within these areas of ambiguity, the EP, Council and the Commission seek to maximise their decision-making authority because 'European institutions ... are always looking for opportunities to enhance their power' (Pierson 1996, p. 133).

On 30 May 1994, the Council confirmed unchanged the common position agreed 11 months earlier. Although the Council members did not accept any EP amendments, most Ministers wanted the EP to adopt the directive and were 'banking on the fact that [the MEPs] would not like to be blamed for the defeat of the legislation, so will accept the common position' (*European Report*, 28 May 1994). Because this was the first failure of a conciliation committee to agree on a joint text, however, the Ministers had little basis for judging whether a majority of MEPs would agree to veto the legislation. Although Thelen and Steinmo (1992, pp. 16–17) argue that political actors are able to adjust their strategies in order to accommodate institutional changes, in this case the actors had difficulty adjusting their strategies, as this was their first experience in this modified institutional setting.

An EP report noted that 'the Council had developed a *legislative culture unused to sharing* political decisions' and conceded that the EP itself 'may well have had *'maximalist' habits* as a result of its history as a largely consultative body'.[43] MEP *rapporteur* Mel Read explains that it was a learning experience for both MEPs and national representatives:

> we were all very unfamiliar with the formal proceedings that were laid down and there were no informal rules about the informal proceedings. So the whole thing was new and cumbersome. I can remember one of the Council members saying, 'Well, why are we talking to the Parliament? What is this about?' The President of the Parliament replied, 'Well you signed the Maastricht Treaty. This is a new power.' It was a learning experience for us all.[44]

As a result of the new institutional authority that the EP gained under the Maastricht Treaty, it was able to reject the Council's common position on 19 July 1994 with an overwhelming 373 votes against, 45 votes in favour and 12 abstentions.[45] This was the first time that the MEPs had used the co-decision procedure to reject a Council text.

This vote turned 'the spotlight on a power struggle between the Council of Ministers and the European Parliament, a fight where each side is determined to have the last word' (*Tech Europe*, 11 July 1994). MEP *rapporteur* Mel

Read was 'amazed that there was no real acknowledgment by Council that the balance of power had changed post-Maastricht'.[46] Karl von Wogau, EMAC chairman, issued a statement:

> When examining the matter, [the EP] expressed its full agreement with the Directive's objectives ... The Council's intransigence did not allow for a compromise to be found in the framework of the procedure of reconciliation. In its common position, the Council did not at all take into account [the EP's] amendments on the substance while major progress had been accomplished during the reconciliation procedure. Moreover, the Council refused to show any sign of flexibility with regard to comitology. For [the EP], it is inadmissible that through comitology, the Council attempt to elude the question of equivalency of treatment throughout the procedure between [the EP] and Council, which are now co-legislators, as stipulated in the Treaty of Maastricht ... For the [EMAC], a new text could be rapidly adopted, considering the progress already accomplished during the reconciliation procedure and the absence of opposition between [the EP] and the [Commission] on the objectives of the directive. Quoted in *Agence Europe*, 29 July 1994.

The EP notes that prior to the EP rejection, 'the Council was unaware of the profound change that the codecision procedure had made to the institutional equilibrium'.[47] Sending out a clear message to the Council to acknowledge the EP's increased authority, Parliament President Klaus Haensch pointed to the failed proposal as 'proof of Parliament's vigilance and the uselessness of attempts not to take account of its stances' (*Agence Europe*, 29 July 1994). The EP veto demonstrates that the new institutional setting has increased the ability of the EP to affect legislation after the Council common position.

This case study also highlights the way in which inconsistent rules and procedures can lead to disputes over institutional design that cannot always be resolved through normal legislative procedures. Whereas the EP and the Council share formal legislative authority in some policy-making procedures, the EP had no formal role in other, related procedures. This contentious aspect of the institutional setting led the EP to reject the voice telephony proposal. In a joint-decision system, where institutional actors seek to maximise their authority, the failure of the proposal demonstrates the importance of clearly defined and well-balanced rules and procedures.

An emerging *modus vivendi*
Having recognised the need to adapt the post-Maastricht policy-making framework, the Member States agreed to enhance the EP's role in committee procedures. Less than six months after the voice telephony proposal was vetoed, the Council, the EP and the Commission signed a *modus vivendi*, under which the EP was to be kept informed of all committee proceedings,

consulted before legislation was modified, and have its recommendations considered. The *modus vivendi* ensured that:

> the Commission will take into account, insofar as possible, any observations by the European Parliament, and will inform the Parliament, at all stages of the procedure, of the action it intends to take, to allow the Parliament to carry out its responsibilities with full knowledge of the issue.[48]

Although the MEPs complained that the agreement was not 'ideal' because 'Parliament had still not been granted equal status in the legislative decision-making procedure', they felt that it 'represented a step forward in Parliament's powers of democratic control'.[49] These new rules further defined the framework governing the inter-institutional bargaining process, helped overcome the inter-institutional conflict experienced previously and facilitated the passage of the Voice Telephony Directive.

5.2(b)   The Second ONP Voice Telephony Proposal

Following Commission President Jacques Santer's pledge to take full account of the EP's recommendations on all rejected proposals (*European Insight*, 16 March 1995), the Commission forwarded the second voice telephony draft Directive to the Council and the EP on 31 March 1995. The Commission considered the new text to be 'as far as possible the text which was to a large extent agreed between the European Parliament and the Council during the Conciliation Meetings' (*Reuter European Community Report*, 25 April 1995).

Due to the increased influence of the EP, the new proposal incorporated half the amendments (7 out of 14) put forth by the EP during its second reading.[50] In addition, at least two amendments aimed at increasing the role of the EP were taken into account (at least in spirit) as a result of the aforementioned *modus vivendi*.[51] On the other hand, several were not taken on board as Commission officials felt that they were already incorporated adequately into the ONP Framework Directive.[52]

First EP reading of second ONP voice telephony directive

On 16 May 1995, the EP delivered an opinion at its first reading of the proposed directive. 'In addition to the climbdown by national governments on the issue of review powers (as a result of the *modus vivendi*)', commented EMAC *rapporteur* Mel Read, 'the new draft had been improved by the Commission on other points of parliamentary concern' (*Euro Watch*, 7 June 1995). She advised MEPs to endorse the proposal because it 'took full account of the parliament's worries', especially concerning transparency and consumer protection (*EuroWatch*, 7 June 1995; *Reuter European Community Report*, 16 May 1995). 'In sharp contrast to the painfully slow, two-year progress of the original draft directive', the EP duly approved the draft proposal without amendment or debate (*EuroWatch*, 7 June 1995).[53]

First Council reading

Demonstrating that the lengthy delay in adopting the proposal had begun to concern market players, ETNO issued a statement declaring the proposal to be out of date considering the EC plans to liberalise the sector.[54] Echoing these concerns, the Council questioned the suitability of the proposal considering the Commission's plans to propose, before 1998, adapting ONP legislation to the new economic environment (*Reuter News Service–Western Europe*, 13 June 1995; *Agence Europe*, 13 June 1995). Three Member States that had recently joined the EU, Austria, Sweden and Finland, also expressed concern over the proposal.[55] As a result, the Council inserted provisions into the proposal that dealt with updating it, and applying it to liberalised markets.[56]

On 12 July 1995, the Telecommunications Council adopted the common position.[57] Portugal was disappointed that the new text did not take into account previous objections to the regulatory committee and voted against the proposal. The Portuguese delegation explained that:

> it is for the Council (acting by co-decision with the European Parliament) to lay down the objectives for convergence laid down in Article 25(2), rather than the Commission as stipulated in this Article. Furthermore, the Portuguese delegation considers that it is totally inexpedient to lay down measures of a compulsory nature for aspects which are not directly linked to the establishment of the internal market whilst making provision at the same time for the adoption of mere recommendations on setting up Europe-wide services requiring Community cooperation. It therefore deplores the failure to reach an agreement more in keeping with the logic of Community law – if only for the sake of consistency ... The Portuguese delegation, moreover, can but deplore the fact that its proposals concerning users' interests, such as omission free of charge, at their request, from telephone directories (an aspect which was also in the Commission's initial proposal), were not accepted.[58]

Before the introduction of qualified majority voting (QMV) under the SEA (see chapter three), Portugal would have had the legislative authority to ensure that its views were taken into account. However, the changed institutional setting reduced the capacity of individual Member States to influence policy outcomes. Under QMV, smaller Member States have even less authority to promote their concerns than the larger Member States: the weighted vote of Portugal counts for half that of the UK, France or Germany. As Scharpf (1997, p. 144) explains:

what has changed is that isolated opponents that cannot claim to be affected in their 'essential national interest' can no longer [block legislation], since if the search for consensus should fail, the majority is now able to have its way by resorting to a vote. As a consequence, the Single European Act has had the effect of greatly reducing the time needed to reach agreement in the Council of Ministers.

The introduction of QMV was an important institutional change, which helps account for the decreased ability of individual Member States to influence policy development.[59]

Once again, however, the two issues that remained unresolved following the conciliation committee, comitology and compulsory compensation, threatened to disrupt the directive's adoption. The Council re-inserted a type IIIa regulatory committee to involve the Member States more closely in decisions taken for the convergence procedure and adjustments made for technical progress. In addition, although the Council acknowledged acceptance of the EP amendment to strengthen compensation provisions,[60] it did not accept the amendment unmodified. In contrast to the EP amendment and the Commission proposal, the wording of the Council's provision no longer provided users with mandatory compensation from TOs in the event of failure to meet contracted service quality levels.

Second EP reading, Commission approval and Council adoption

Even though the Council common position, in combination with the aforementioned *modus vivendi*, took into account (at least in spirit) the same nine out of 14 EP amendments as the Commission proposal (see above), the MEPs were not satisfied. Following an initial reading and discussion over the common position, EMAC adopted two amendments due to the resurfacing of concerns over mandatory compensation and comitology.[61]

However, perhaps recognising the lengthy delay that the proposal had already faced, the EP and Council were finally able to resolve all of their differences and come to agreement on a joint text. Following an informal consultation between *rapporteur* Mel Read and a Council representative, the EP approved two 'watered down' amendments in October 1995.[62] EP acceptance of these two amendments demonstrates, once again, the importance of both the EMAC and these informal inter-institutional meetings during the inter-institutional bargaining stage.

Following Commission approval of the two amendments, the Council adopted, without discussion, the amended ONP Voice Telephony Directive on 29 November 1995.[63] The significant delay the directive faced is exemplified by the fact that its adoption took place more than five years after the Commission first commissioned studies on the subject and more than three years after the Commission forwarded its first draft proposal to the EP and Council.

5.3 CONCLUSION

This chapter confirms Peterson's assertion that new institutionalism is the best theory for analysis of the EC's systemic level (Peterson, 1995a). In particular, rational choice institutionalism highlights the fact that MEPs, Commission officials and Council members are strategic, self-interested actors, whose actions are channelled through a complex set of EC treaty rules and procedures. At the same time, the approach emphasises that these actors seek to influence the design of the EC's institutions in ways that maximise their own personal utility and policy-making authority. The framework nature of the EC treaty enhances this inter-institutional dynamic; a lot of room is left for self-serving interpretations of their respective competences.

Successive procedural modifications to the inter-institutional bargaining stage have increased the EP's ability to influence telecommunications policy outcomes. Although the cooperation procedure enabled the EP to make important changes to the ONP Framework Directive in its first reading, none of its second reading amendments were incorporated into the directive. The EP gained agenda-setting authority under the co-decision procedure, empowering it to influence the ONP Voice Telephony Directive in ways that were not entirely in accordance with the collective preferences of the Member States. Although the Council initially rejected all EP amendments, following the EP veto of the initial proposal, the Council and Commission adopted a text which, in conjunction with the *modus vivendi*, incorporated a majority of the EP amendments (9 out of 14). The EP's own quantitative analysis of the 63 successful conciliation procedures completed between November 1993 and April 1999 concludes that nearly three-quarters (74 per cent) of the EP's second reading amendments were accepted either wholly or partly.[64] In contrast to accounts of European telecommunications prior to the 1990s, and the examination put forth in chapter three, the EP must therefore now be considered an important institutional actor in the European telecommunications sector, as a result of changes in the inter-institutional balance of power. At the same time, it is likely that the EP has also become an important institutional actor in other policy areas, because the co-decision procedure is increasingly being used in Community decision making (Maurer 2003, p. 231).

This chapter also highlighted that fact that disputes over institutional design are a salient feature of the EC's systemic level and that more clearly defined rules and procedures help to reduce conflict. The inter-institutional agreement on a *modus vivendi*, which gave the EP more input into committee procedures, helped to ease the passage of the second ONP Voice Telephony Directive. At the same time, these rational actors have adjusted their strategies in response to modifications to the institutional setting. The 1994 EP rejection of the ONP voice telephony occurred in one of the earliest uses of the co-decision procedure. Since then, the EP and the Council have worked together more closely and have been more cooperative and remarkably successful at reaching agreements. Of the 417 co-decision procedures completed as of July 2002, only one other (in 1995) has resulted in an EP rejection,[65] while three others have failed because of an unsuccessful conciliation (Maurer 2003). MEP Read commented that the co-decision procedure has forced the EC institutions to work together 'more seriously and openly'.[66] EP vice-president Nicole Fontaine commented, 'the willingness of the two sides to reach an agreement is increasing'.[67] In response to institutional change, political actors have thus undergone a 'process of mutual understanding, adjustment and [the] *gradual establishment of parity*.'[68]

Whereas chapter one concluded that the distinction between Peterson's sub-systemic and systemic levels of analysis is unclear (Peterson, 1995a), Part II of this book finds that, in relying entirely on organisational and structural factors, a clear distinction can be drawn between the levels of analysis. At the sub-systemic level, almost all of the rules and procedures governing telecommunications policy formulation are sector-specific. At the systemic level, by contrast, the formal rules governing inter-institutional bargaining are set out in the EC Treaty and generally applicable. Each level of analysis thus has a distinctive institutional setting, which has an important and influential effect on policy development. In addition, although MEP committees are dominant actors in Peterson's sub-systemic level of analysis, this chapter found that the EMAC also has significant influence at the EC's systemic level.

NOTES

1. European Parliament, PE 230.998 'Activity report 1 November 1993–30 April 1999 on the Codecision procedure', p. 4.
2. Following the Treaty of Amsterdam, the cooperation procedure applies exclusively to the field of economic and monetary union (Articles 99(5) and 106(2) of the EC Treaty).
3. In addition, the COR was established in 1994 and thus was not significantly involved with the proposals examined in this chapter.
4. European Commission, *ONP Framework Directive*, COM(89) 825 final – SYN 187 of 9 January 1989, OJ C 39, 16.2.1989, p. 8.
5. European Parliament, *Report on draft ONP Framework Directive*, Doc. A 2-122/89, PE 132.684; OJ C 158, 26.5.1989

6. Interview, MEP Fernand Herman, Brussels, April 1999; Interview, MEP Mel Read, telephone, July 2000; Interview, ETNO representative, Brussels, May 1999; Interview, DG Information Society (DG XIII) official, Brussels, May 1999.
7. Interview, MEP Fernand Herman, Brussels, April 1999; Interview, MEP Mel Read, telephone, July 2000.
8. Interview, Brussels, April 1999.
9. Interview, MEP Fernand Herman, Brussels, April 1999.
10. Interview, MEP Fernand Herman, Brussels, April 1999.
11. European Commission, *ONP Framework Directive*, COM(89) 325 final – SYN 187 of 10 August 1989.
12. Interview, DG Information Society (DG XIII) official, Brussels, April 1999.
13. The difficult political compromise of December 1989 and, particularly, its relation to the Services Directive, will be discussed further in chapter seven.
14. European Parliament, Council explanatory document cited in PE 139.188 'EMAC Second Reading opinion'.
15. European Parliament, Council explanatory document cited in PE 139.188 'EMAC Second Reading opinion'.
16. European Parliament, PE 139.188 'EMAC Second Reading opinion'.
17. These time limits may, however, be extended for a limited period through an agreement between the EP and Council.
18. Hoping to address what were perceived to be the most important issues left out of the Council's common position, the amendments sought to protect data (No. 25 in Table 5.2), ensure the financial viability of the network operators (No. 37) and prevent ONP conditions from restricting the interoperability of services. European Parliament decision, doc. A3-85/90, OJ C 149/82, 18.6.90.
19. Interview, DG Information Society (DG XIII) official, Brussels, March 1999.
20. Interview, Former DG Information Society (DG XIII) official, Brussels, July 1999.
21. Interview, DG Information Society (DG XIII) official, Brussels, March 1999. Commission officials were particularly concerned in the light of the Advocate's General opinion of February 1990 that the Article 90 legal basis of the Terminal Equipment Directive had been incorrect (see chapter seven). Commission officials felt that this opinion weakened the Article 90 legal basis of the Services Directive and, thus, could undermine the December 1989 political compromise.
22. Six amendments in the first reading and an additional amendment in the second reading were not accepted.
23. For an overview of the European Parliament's authority under the co-decision procedure and a more detailed description of the changes made to the procedure after the Amsterdam Treaty, see Maurer (2003, especially pp. 228–230).
24. European Parliament legislative Resolution, Doc. A3-0064/93, OJ C 115/105, 26.4.93.
25. As chapter 4 discussed, the Commission's proposal provided for a type I advisory committee, under which the Commission maintains autonomous control and the Member States can only put forth non-binding advisory opinions.
26. Interview, First Secretary Responsible for Telecommunications, UK Permanent Representative, Brussels, May 1999.
27. For a discussion of the importance of early lobbying in the EC policy-making process, see Hull (1993).
28. EP legislative Resolution, Doc. A3-0006/94, OJ C 44/93, 14.02.1994.
29. Amendments 4, 5, 6, 10, 15, 17, 21, 23, 29, 30, 34 and 35 (see Table 5.4) were reintroduced. The most important sought to enforce EP and industry consultation (nos 5, 6, 30, 34 and 35); protect small and medium-sized firms (nos 10, 21, and 23); reinforce compensation provisions (no 15); and require Commission notification for service access restrictions (no 29).
30. European Parliament, PE 206.718/fin.
31. DEP 3-440, 14 December 1993, p. 94 (quoted in Earnshaw and Judge 1996, p. 116).
32. European Parliament, PE 206.718/fin.

130 *The harmonisation of European telecommunications policies*

33. This included the amendments designed to safeguard small telecommunications firms (nos 10, 21 and 23) as well as those dealing with the consultation arrangements between concerned industry groups (nos 5, 30 and 35). EP Session News Press Release, 26 October 1993.
34. This included Amendment numbers 6 and 34 and the new amendment eliminating reference to a type IIIa regulatory committee. *European Report*, 30 March 1994
35. These concerned Commission notification for the denial of access to the public telephone network (no. 29); 'prompt and justified' explanations for the denial of special network access (no. 17); compulsory compensation for the failure to meet service quality levels (no. 15); and the new amendment to require mandatory auditing of TO accounts (*Reuter European Community Report*, 17 March 1994).
36. See Rule 75 of the European Parliament Rules of Procedure.
37. Interview, Brussels, April 1999.
38. The Council was flexible on the amendments concerning Commission notification in the case of network access restrictions (no. 29) and the new amendment on the external inspection of TO accounts, while the European Parliament was flexible on the amendments concerning consultation with the telecoms industry (nos 5, 6, 30 and 35), the protection of small and medium-sized TOs (nos 10, 21 and 23) and user rights (nos 4 and 30). *European Report*, 30 March 1994.
39. Amendment numbers 6 and 34 and the new comitology amendment.
40. IIA agreement, 'The Arrangements for the proceedings of the Conciliation Committee under Article 189B', OJ C 329/141, 1993.
41. Council Decision of 13 July 1987 laying down the procedures for the exercise of implementing powers conferred on the Commission (87/373/EEC), *Official Journal* No L 197/33.
42. The document *European Union* (Office for the Official Publications of the European Communities) also states that the Council must reach a unanimous decision if its recommendation differs from the Commission proposal (Nugent 1994, p. 317).
43. European Parliament, PE 230.998 'Activity report 1 November 1993–30 April 1999 on the Codecision procedure', p. 5, italics in original.
44. Interview, telephone, July 2000.
45. European Parliament, EP legislative Resolution, Doc. A4–0001/94.
46. EP Session News Press Release 'Telephone services and open network provision – MEPs reject text (A4–1/94)' 19 July 1994 (quoted in *Reuter EC Report* 22 July 1994).
47. European Parliament, PE 230.998 'Activity report 1 November 1993–30 April 1999 on the Codecision procedure', p. 6.
48. '*Modus vivendi* between the European Parliament, the Council and the Commission concerning measures to implement legislation adopted according to the procedure set out in Article 189 of the EU Treaty', cited in EP Session News Press Release, 18 January 1995.
49. EP Session News Press Release, 18 January 1995.
50. In particular, the Commission incorporated into the new proposal the amendments aimed at increasing openness and transparency (nos 5 and 6); and those establishing requirements for compulsory compensation (no. 15); prompt and justified explanations for the denial of special network access (no. 17); Commission notification in the case of network access restrictions (no. 29); the mandatory auditing of TO accounts (at least in spirit); and the removal of the reference to a type IIIa regulatory committee.
51. Amendments 34 and 35.
52. The Commission did not incorporate the EP's amendments aimed at ensuring the establishment of effective NRAs (nos 4 and 30) and those aimed at protecting small and medium-sized TOs (nos 10, 21 and 23) (European Parliament, EP legislative Resolution, Doc. A3-0006/94, OJ C 44/93, 14.02.1994; European Commission, Commission proposal for an ONP voice telephony directive, COM (94) 689 final- 95/0020 (COD) 95/C 122/04, 18.5.95).
53. European Parliament, EP legislative Resolution, Doc. A4-0090/95, OJ C 151/27, 19.06.1995.

54. ETNO explained that 'the directive does not reflect the fact that the regulatory context for networks and public voice telephony is already evolving and will change by 1998 ... [therefore] the Commission should propose a new text that adapts the ONP concept to liberalised markets ... [because] by the time the current text is implemented, it would only be valid for about a year before the 1998 liberalisation deadline kicks in' (quoted in *Reuter News Service – Western Europe*, 31 May 1995).
55. Austria expressed concern over the whole text, while Sweden and Finland (joined by the UK), having already liberalised extensively, questioned the application of the directive to countries that had already dropped exclusive rights for supplies of voice telephony services (*European Report*, 19 April 1995).
56. Council of Ministers, Commission Position on ONP Voice Telephony Directive; Article 32(1) called on the Commission to propose a revision of the directive 'to the requirements of market liberalization' and Article 26(2) provided a description of the implementation of the directive in Member States that had previously liberalised voice telephony services. See Common Position (EC) No 17/95 (1995 OJ C 281).
57. Common Position (EC) No 17/95 (1995 OJ C 281).
58. Council of Ministers Press Release, 13 June 1995.
59. Another factor that helps account for the reduced ability of individual Member States to influence policy outcomes is the increasing number of Member States. In 2004, ten countries joined the European Union, bringing the membership total up to 25 Member States. The rising number of Member States tends towards diluting the influence of each Member State individually.
60. The Council noted that the common position 'specifies that any compensation and/ or refund arrangements to be provided if service quality levels are not met must be included in user contracts. The European Parliament and Council were divided on this question during the co-decision procedure on the previous proposal, and the latter has therefore agreed to take the European Parliament's concerns as to consumer protection into account.' Common Position (EC) No 17/95 (1995 OJ C 281).
61. The first amendment required the NRAs to 'ensure' compensation for failure to meet contracted service quality standards. The second amendment recalled the *modus vivendi* agreed upon as a temporary resolution to the comitology dispute (*European Report*, 30 September 1995).
62. The first amendment required TOs 'as a general rule' to compensate customers if contracted service quality standards were not met. Any exceptions to this rule were to be made clear in the users' contract and required justification. The second amendment added a reference to the *modus vivendi* in the preamble of the directive. European Parliament EP legislative Resolution on ONP Voice Telephony Directive, Doc. A4-0231/95, OJ C 308/112, 20.11.1995.
63. Once again, Portugal voted against the directive and Spain abstained. ONP Voice Telephony Directive, Directive 95/62/EC (1995 OJ L 321).
64. European Parliament, PE 230.998 'Activity report 1 November 1993–30 April 1999 on the codecision procedure', p. 11.
65. European Parliament, PE 230.998 'Activity report 1 November 1993–30 April 1999 on the codecision procedure', p. 48. The only other rejection of a Council text by the EP (even though the conciliation committee had agreed upon a joint text) concerned biotechnology in March 1995.
66. Quoted in the *Reuter European Community Report*, 6 February 1996.
67. Quoted in the *Reuter European Community Report*, 22 February 1995.
68. European Parliament, PE 230.998 'Activity report 1 November 1993–30 April 1999 on the Codecision procedure', p. 5, italics in original.

6. Limits to the Europeanisation of telecommunications

With the benefit of hindsight it would now seem unrealistic to expect that where harmonisation cannot be achieved through the Community's supranational decision making procedures, this should be feasible through the intergovernmental framework of CEPT.[1] *Commission draft directive on electronic authorisations, July 2000.*

INTRODUCTION

This chapter continues to examine the capacity of various national and European actors to re-regulate or harmonise national telecommunications legislation. Whereas the previous chapters concluded that the Commission and the EP are important institutional actors and have some agenda-setting capabilities with respect to the harmonisation of European telecommunications policies, this chapter nonetheless argues that national government officials can have, in defending common interests (for example, ensuring subsidiarity for actions better handled at the national level), an even greater influence over policy development. Moreover, this increased authority largely results from their privileged position in the institutional environment.

This chapter employs the policy networks approach to help evaluate the influence of competing groups of actors. A large network of institutional and political actors, including Commission officials, MEPs and industry and user representatives, favoured the establishment of a European Regulatory Authority (ERA) and the mutual recognition of telecommunications licences. Ultimately, two factors prevented these actors from achieving their preferred policies: 1) the resistance of the opposing intergovernmental policy community of national policy makers and regulators, notably those from the UK, France and Germany, and 2) the institutional opportunity structure.

This chapter also evaluates Fritz Scharpf's (1988) contention that the EC policy-making process suffers from a joint-decision trap. As discussed earlier, this arises because EC decision making is directly dependent upon the agreement of the Member States, and this agreement must be either unanimous or nearly unanimous. According to Scharpf (1988, p. 254), the joint-decision trap contributes to the 'substantive deficiencies of joint policy making in the European Community'. This chapter finds that the joint-decision trap helps account for the EU's difficulties in establishing an ERA and an EC-wide licensing regime. The existence, organisational design and resistance of the intergovernmental CEPT framework further hindered the development of a coherent set of harmonised licensing conditions. As a result, even following implementation of the Licensing Directive in 1998, Member States maintained widely divergent national approaches to the licensing of telecommunications operators.

6.1   INSTITUTIONAL BARRIERS TO A EUROPEAN REGULATORY AUTHORITY

Consistent with the trend, since the early 1990s, of establishing regulatory agencies at the European level (Kelemen 2002),[2] industry, users, the EP and the Commission favoured the establishment of a telecommunications regulator with pan-European authority. In seeking to achieve this objective, however, these actors failed to form an influential policy network because they lacked sufficient policy-making resources. As Peterson (1995b, p. 403) points out, defining the boundaries of a policy network is, theoretically, 'relatively straightforward', by 'determining which actors or institutions possess resources sufficient to affect policy outcomes'. By logical extension, therefore, one of the preconditions for the establishment of a policy network is that institutions or actors must have sufficient resources, whether 'constitutional-legal, organizational, financial, political or informational', to influence policy output (Rhodes 1997, p. 37).

The group of actors that favoured an ERA, however, did not maintain sufficient resources; in particular, and as discussed further below, they lacked formal policy-making authority to coalesce into a powerful policy network on this issue and have been unable to achieve their desired policy line. National policy makers and national regulatory authorities, on the other hand, have maintained a powerful policy community and have been able to resist any such transfer of competence to the European level. The institutional opportunity structure has, on this issue, favoured the Member States.

Widespread Support for an ERA

Key political and institutional actors supported the establishment of a regulatory authority at the European level. There was widespread support among users and industry, from the late 1980s until the mid- to late-1990s, for the establishment of an ERA with powers similar to OFTEL in the UK.[3] User groups, including the International Telecommunications Users Group (INTUG) and the European Council of Telecommunications Users Associations (ECTUA), had argued, as early as 1989, for the need to establish a European Regulatory Authority (ERA).[4] Users were primarily concerned that market distortion would result from differences in the national implementation of the ONP legislation.

In 1994, the European Space Agency submitted a document to the Commission, which summarised the findings of consultations with more than 150 representatives of the satellite communications industry.[5] The document highlighted 'the industry's plea to establish a European regulatory body,'[6] and found that '... *there is strong support for an independent and efficient European regulatory authority (EURO-OFTEL or OFSAT) to ensure fair play*'.[7] Similarly, following consultations on a 1994 Green Paper to liberalise mobile communications, the Commission concluded that 'a number of contributors ... argued strongly for a single regulatory agency at a Union level which could manage common resources (such as frequency and numbering) and which could ensure a consistent approach in national implementation of the Union regulatory framework'.[8]

The EP also advocated the establishment an ERA at the Community level.[9] The EP proposed amendments to directives in 1996 and 1997 to include the option of establishing an ERA for certain oversight functions.[10] 'It is time to give serious consideration to the idea of [an ERA]', argued MEP *rapporteur* Mel Read at a 1996 EP Plenary session (*Agence Europe*, 20 February 1996). Although the Council refused to accept these amendments, the EP used its authority under the co-decision procedure to force the Council to accept a provision that called on the Commission to report on the possibility of establishing an ERA.

In line with the institutional mission of the Commission to seek supranational solutions to coordination problems, both the Telecommunications and Competition Directorates-General (DGs) favoured the establishment of an ERA.[11] In December 1993, Michel Carpentier, Director-General of Telecommunications, argued that 'issues such as the resolution of interconnection disputes between telecommunications organisations in different member states and the mutual recognition of licenses cannot by their nature be handled by one member state or one regulator acting alone'.[12] In a May 1994 Report to the European Council, Telecommunications Commissioner Martin Bangemann recommended that 'an authority should be established at the European level whose terms of reference will require prompt attention'.[13]

In a submission to the December 1994 European Council, the Commission noted that the multiplicity of national licences would hamper the development of trans-European services at the national level. 'Ultimately, it might be worth considering a European authority capable of resolving this type of problem.'[14] In July 1996, Competition Commissioner van Miert explained that:

> there are of course already many existing coordination bodies drawing together national regulations and fostering cooperation ... but however many there are it is becoming increasingly apparent that these are not sufficient. The main point I want to make is that we should focus upon exactly and only those areas where it is truly called for. In order to be a viable and workable reality [an ERA] should have a mandate of clearly defined and mainly technical tasks such as numbering and spectrum management.[15]

As late as 1998, Telecommunications Director-General Robert Verrue spoke of the benefits of an ERA.[16] Although Commission officials, MEPs, industry and user representatives favoured the establishment of an ERA, they failed to organise significantly, or develop adequate 'linkages' (Kenis and Schneider 1991, p. 41) at the EC level, in pursuit of this objective. In this case, these actors lacked the authority to get the issue of an ERA actively considered at the EC level. This is crucial, because 'no policy can be made if the issue to which it is addressed cannot first be placed onto the active agenda' (Peters 1996, p. 61). The reasons for this failure are due largely to the institutional framework, discussed below, which helped shape the politics surrounding discussions of an ERA.

The Joint-Decision Trap

Two important institutional considerations have hindered the development of a significant EC-level debate over the establishment of an ERA. First, this would require unanimous Member State approval. To establish an ERA with wide-ranging powers, including policy functions and discretion in its activities, an amendment to the EC Treaty would be required. If an ERA were to be defined with narrower, more circumscribed executive powers and clearly defined activities, it would be possible to use Article 308 of the EC Treaty (formerly Article 235).[17] In both cases, however, the establishment of an ERA requires unanimous Member State approval. This institutional setting thus strongly favours national interests; any of the 15 Member States can veto the establishment of a European level regulatory body. It is argued here that this institutional framework, identified as the joint-decision trap, has enabled the Member States to resist the development of an ERA.

Second, the problem of the joint-decision trap was further compounded by the Community's normative shift towards respecting subsidiarity in the early 1990s. The subsidiarity principle implies that the Commission should not assume responsibility for any task that the Member States perceive to be a national responsibility. Moreover, following the ratification of the Maastricht Treaty, the rules of the game changed in accordance with this principle.[18] All potential EC legislation is required to satisfy the so-called subsidiarity test. Worthy and Kariyawasam (1998) argue that it is unlikely that the establishment of an ERA would be able to satisfy this test.[19] The EC's commitment to subsidiarity therefore provided a further hindrance to the establishment of an ERA.

The Member States were effectively able to keep the issue of an ERA off the formal agenda because of the joint-decision trap and the normative commitment to subsidiarity. Peters (1996, p. 61) describes this early part of the policy-making process as 'an initial crucial veto point', where 'political and administrative leaders can exercise their power ... to prevent anything from happening'. In fact, the Telecommunications Council never even discussed the establishment of an ERA formally. Although *The Economist* asserted that the November 1994 Telecommunications Council was likely to have been the best opportunity to discuss an ERA, the 'telecoms ministers ... dodged the issue'.[20] In response to a question regarding the possible creation of an ERA from MEP Ludivina Garcia Aria (PES, E) in July 1995, Spanish Telecommunications Minister and Council President Jose Borrell noted, 'the question could be envisaged if a concrete proposal were made. At any rate, one should take the principle of subsidiarity into account' (*Agence Europe*, 25 July 1995). Nonetheless, it was Member State opposition, in conjunction with the subsidiarity principle, which kept the Commission from proposing an ERA to the Council formally.

Faced with this resistance, the Commission adopted the same strategy it employed during previous attempts at expanding its telecommunications authority (see chapters 3 and 4); it enlisted the services of an epistemic community to garner support for its desired policy line. In 1997, the Commission thus funded a study, *Issues associated with the creation of a European regulatory authority for telecommunications*, by the National Economic Research Associates (NERA). NERA's study was based on 50 interviews with key market players and policy makers, carried out during the summer of 1996 in nine Member States: Austria, France, Germany, Greece, Italy, the Netherlands, Spain, Sweden and the United Kingdom. Opinions of interviewees on the establishment of an ERA are summarised in Table 6.1 and more specifically, on the establishment of an ERA with licensing responsibilities in Table 6.2.

These two tables demonstrate that a large majority of market players felt that an ERA would have been beneficial for the EC regulatory framework. National policy makers and regulators were the only actors clearly opposed to the idea. The survey found that 'regulators and policy makers were mostly against the creation of a new European regulatory body ... while the vast majority of competitors, users and other interviewees considered that there might be at least some advantages in the creation of a new body dealing with some aspects of telecommunications regulation at the level of the European Union'.

Table 6.1   Survey: Would an ERA add value to the current regulatory framework within the EU?

Interviewee Type	YES	NO	DON'T KNOW
Principal TOs	4	2	2
Regulators and Policy Makers	3	6	1
Competitors	18	3	2
Users	6	1	
Member of European Parliament	1		
Equipment Manufacturer	1		
Total	33	12	5

Source: NERA (1997), p. 21.

Table 6.2   Survey: Should an ERA have a role in licensing?

Interviewee Type	YES	NO	DON'T KNOW
Principal TOs	2	2	4
Regulators and Policy Makers	1	8	1
Competitors	12	5	6
Users	5	1	1
Member of European Parliament	1		
Equipment Manufacturer			1
Total	21	16	13

Source: NERA (1997), p. 30.

Aware that an ERA would require the support of the national government representatives, Commission officials refrained from proposing an ERA formally. The prevailing institutional setting, which privileged Member State representatives over Commission officials and market players, thus served as a significant barrier to the formation of an ERA. In particular, the normative commitment to subsidiarity, along with conflicting policy frames and interests, helped shape the politics surrounding ERA discussions. The NERA study concluded that the 'opponents of a new European Regulatory Authority tended to feel that the creation of such a body would conflict with the principle of subsidiarity'.

The historical institutionalist concept of path dependency helps to account for some Member State opposition to an ERA. As chapter two explained, each of the Member States had established a national regulatory authority by the early 1990s. Since these governments had recently expended significant resources in constructing their own national regulators, each with different degrees of power according to national preferences, they were not keen to abandon them; the Member States wanted to proceed down the path they had previously chosen and to ensure the continued development of these newly established institutions. The Member States were afraid that, if the decision were made to form a European regulator, their regulatory authorities at the national level would be made redundant.

In line with Pierson's (1996, p. 145) claim of 'path dependency' or the 'continued movement down a specific path once initial steps are taken', national policy makers argued that the newly established NRAs would be sufficient to meet harmonisation needs. The NERA report (1997) noted that the opponents felt that 'current institutions and procedures, which could continue on a path of development and improvement, would suffice in providing the measures required for the development of a single market'. OFTEL (1996), for example, wished to preserve its powers and argued that the 'machinery and powers are in place – through a EU framework implemented on a day-to-day basis by NRAs – to deliver a competitive EU market. A new "Euroregulator" body does not seem to be justified.'

Institutions were thus important in structuring the politics surrounding the discussions of an ERA and helped determine the policy outcome; national governments were able to prevent a new EC-level institution from evolving due to their privileged position in the EC institutional framework. While the advantages of a single telecommunications market had spurred the Member States to agree on some harmonisation, its 'attraction was certainly not sufficient to persuade national governments to commit institutional suicide' (Scharpf 1988, p. 243). In other words, the Member States were not about to abandon their national regulatory authorities; instead of transferring further competencies to the EC level, the national governments hoped to maintain control over the regulation of cross-border telecommunications. As Eberlein and Grande (2005, p. 104) have similarly concluded, 'divergent interests of members states run counter to the goal of delegating regulatory competencies to the European level, thus the chances of 'positive' re-regulation of negative market integration are limited ... As a consequence, the EU's formal powers and its institutional capacities have so far remained underdeveloped.'

6.2 MEMBER STATE OPPOSITION TO LICENSING HARMONISATION

This section puts forth two arguments. First, with sufficient resources, a group of actors can form an influential policy network to promote a common interest. In this case, the Commission, with the support of the EP, service providers and users, favoured the establishment of an EC-level licensing regime. In contrast to the above discussion of an ERA, proposals for an EC licensing regime were put forth under the co-decision procedure, thus granting both the Commission and the EP greater formal policy-making authority (see chapters four and five). Service providers and users were mobilised into action through the institutional setting established by the Commission in the early 1990s for the formulation of telecommunications legislation (see chapter four). As a result, this group functioned as a stable policy network, able to promote their shared interest in harmonised licences.

Second, notwithstanding the significant influence that the Commission, the EP, service providers and users maintain over the formulation of EC telecommunications policies, the EC policy-making process favours the Member States. Although the Commission was able to use its formal agenda-setting authority under the co-decision procedure to get its licensing proposals considered, because the approval of a qualified majority of Member States is required, national policy makers and regulators were able to water down the Commission's initial harmonisation intent. Thus, agenda-setting capacity does not guarantee policy influence.

Once again, the national representatives were well organised through the Council of Ministers and sought to prevent the further transfer of competencies to the supranational EC. With the support of most former PTTs, this intergovernmental policy community wanted to retain as much influence as possible in the harmonisation of European licensing conditions. Their position was enhanced by the inter-institutional debate surrounding subsidiarity following the Maastricht ratification process.[21]

The Failed Single Telecommunications Licence and Mutual Recognition Approach

Given the widely differing licensing regimes in the Member States, the Commission proposed a directive on the mutual recognition of licences in September 1992.[22] In line with the policy formulation stage outlined in chapter four, the proposal was based on substantial studies produced by an epistemic community of external consultants, institutionalised and pluralistic consultations with interested parties and several discussions between the Commission and the ONP committee.[23] The proposal sought to establish a single EC telecommunications licence, which would have enabled the holder to provide services throughout the EC.

The proposal also would have established a Community telecommunications committee (CTC), composed of NRA representatives, to advise the Commission.[24] The CTC would have been able to determine which types of services the telecommunications operators could provide without a licence or prior authorisation. Through the CTC, the Commission sought to incorporate further the NRAs into the policy-making process in order to institutionalise licensing authority at the EC level. Because licensing had previously been the sole domain of the NRAs, the Commission recognised the need to give the powerful NRAs a key role in order to increase its own authority over the sector.[25]

With the support of the EP, service providers and users, Commission officials argued that having service providers apply for separate licences for each Member State in which they wanted to operate was 'cumbersome' and 'time-consuming' and restricted the provision of EC-wide telecommunications services (*Tech Europe*, 10 September 1992). Once again, however, the institutional opportunity structure favoured the opposing coalition of national policy makers, telecommunications organisations and regulatory authorities.

> The concept [of mutually agreed licences] was greeted by users and service providers with general enthusiasm as a necessary step on the path towards pan-European networks ... Opposition, however, has been forceful – both from Member States jealous of the independent powers of their National Regulatory Authorities, and from TOs uncertain or unwilling to assist any moves away from the national monopoly environment. (Analysys Consultants 1992, p. 60)

In particular, the intergovernmental policy community of national regulators and policy makers expressed concern over the balance of authority in the draft proposal (*Communications Week International*, 8 March 1993, p. 3). The Member States saw the proposed CTC as an attempt to establish the forerunner to an ERA. In line with the conclusions of successive European Councils, and in the spirit of the Maastricht Treaty, the French government submitted a Memorandum to the May 1993 Telecommunications Council calling on the Commission to respect the normative commitment to subsidiarity.

Le principe de subsidiarité doit être en particulier le fondement d'une répartition harmonieuse des rôles entre Communauté et Etats membres. L'application des dispositions communautaires doit être ainsi confiée prioritairement aux autorités de réglementation nationales, notamment pour la gestion des problèmes réglementaires (licenses, numération, fréquences, tarifs); la nécessaire coopération pour résoudre les questions de dimension pan-européenne doit s'appuyer autant que possible sur la CEPT.[26]

Confirming the close link between institutional norms and formal rules and procedures (March and Olsen 1989), France once again invoked the principle of subsidiarity to justify a particular institutional design that would limit Commission authority over licensing harmonisation. French authorities preferred to maintain a more autonomous role for the NRAs and argued that the intergovernmental CEPT was the most appropriate organisation to pursue the harmonisation of licenses. As pointed out in chapter three, France was the Member State that had pushed most strongly for the establishment of the CEPT in 1959, as French officials wanted to coordinate European telecommunications policies outside the supranational EC. This preference for national autonomy, and a more limited role for the EC, continues to be influenced by the traditional French *étatist* policy style examined in chapter two. The Commission was forced to abandon the draft directive. A Commission official who worked on the proposal noted that 'we weren't close ... it was shut down immediately by the Member States'.[27]

Since the draft directive did not apply to satellites, the Commission also adopted a separate proposal in January 1994 for a directive on the mutual recognition of licences for satellite communications services.[28] Commission officials noted that a multinational company, which wanted to develop an internal satellite communications network, needed to complete as many as 5000 separate applications for licences (*Tech Europe*, 4 February 1994). However, the Member States resisted this approach as well. The Commission noted that:

> One source of concern is that most of the Member States are reluctant to agree to *mutual recognition of licenses* for satellite networks services and satellite communication services. Maintaining a multiplicity of national licenses will obviously hamper the development of services at trans-European level.[29]

As *TechEurope* (1 December 1995) noted, 'the EU Member States have always opposed any attempt to get them to relinquish their prerogatives as regards licensing'. Commission officials, users and service providers were forced to abandon their approach. The co-decision procedure thus, in this case, did not provide this network with sufficient authority to achieve their preferred policy line of a single telecommunications licence or the mutual recognition of licences. In contrast to previous chapters, where Commission officials and users were found to be influential, the Member States here were more unified in their opposition to EC-level licensing coordination.

A Watered Down Approach

Even though the Member States rebuked their initial attempt, Commission officials and users came forward with a new approach for harmonising licensing conditions. In November 1995, the Commission adopted a proposal for a watered down Licensing Directive.[30] This draft directive replaced the two mutual recognition proposals and, 'in line with the principle of subsidiarity, the granting of authorisations remain(ed) the responsibility of Member States'.[31] In place of the failed mutual recognition approach, the Commission, with the support of users and service providers, resorted to soft law and other policy instruments to obtain their objectives.

In particular, the draft directive set out five principles that the NRAs were required to follow in licensing telecommunications operators. First, national licensing and authorisation conditions were to be 'objectively justified, … non-discriminatory, proportionate and transparent'. Second, priority was to be given to general authorisations under which all operators (that comply with general rules) could offer services without submitting an application. Third, individual licences were to be required only in certain circumstances (for example, radio frequencies, numbers, land access or public service obligations). Fourth, there were to be an unlimited number of new licences (except to ensure that radio frequencies are used efficiently). Fifth, licence fees were only to cover administrative costs (although a charge to cover the value of scarce resources, such as radio frequencies, could be imposed).

These five principles nonetheless gave the NRAs freedom to establish dissimilar licensing regimes. NRAs could attach conditions to general authorisations; require individual licences in certain situations; and impose fees higher than costs in limited circumstances. National government officials were thus able to use their privileged position in the institutional setting to maintain a degree of national autonomy over licensing procedures.

In line with French demands, harmonisation efforts were to remain the responsibility of the intergovernmental CEPT instead of the supranational EC, thus necessitating Member State consensus to move forward. The Commission was required to award mandates to the CEPT for the harmonisation of licensing conditions, procedures and fees. In addition, in place of the Commission's mutual recognition approach, the proposal also mandated the CEPT to establish a one-stop shopping agreement so that European operators could at least obtain their distinct national licences from one organisation. The Commission noted that:

> The new approach will retain one important element of our earlier mutual recognition approach by using [the CEPT] to develop specific harmonised license conditions. Pending agreement on harmonised licenses, the new proposal would institute a one-stop shopping procedure for individual licenses with set time limits which help businesses to provide or to use new Europe-wide services.[32]

As discussed further below, however, the organisational features of the CEPT, combined with the continued resistance of the NRAs, further hindered the harmonisation of licensing conditions and prevented the establishment of the one-stop shopping procedure.

Inter-Institutional Bargaining

Although the Commission adopted approximately two-thirds of the EP's first reading amendments (26 out of 37) in its amended proposal,[33] the Council included just over a third of the EP amendments into its common position (14 out of 37).[34] Largely because of its increased influence during the second reading stage (as examined in the previous chapter), however, the EP was able to get the Council to accept all eight of its second reading amendments.[35] The most important of these amendments initiated a Commission investigation into the possibility of an ERA, and introduced a procedure enabling potential competitive operators to appeal to an independent body in the event of an NRA refusal to issue a licence.[36]

The Council nonetheless amended the five principles introduced above to increase the ability of the NRAs to control market access. The Council aimed 'to introduce greater flexibility into the text to take account, firstly, of the wide range of situations, and, secondly, of the complexity of the work of the national authorities'.[37] France, with the support of Greece, Italy, Portugal, Belgium and Spain, sought to give their regulators greater freedom in establishing licensing regimes. The approach of these national governments towards EC-level negotiations is thus consistent with their traditionally strong state role in the sector, including the lingering influence of French *dirigisme* (see second chapter). As Vincent Wright (1997, p. 151) points out, however, the nature of French *dirigisme* had changed: 'the emphasis of *dirigisme* ... shifted from control and direction to guidance and support in an attempt to forge a market-oriented "competitive state"'.

The UK, Finland, Sweden and the Netherlands, on the other hand, had already introduced widespread competition into their markets and felt that 'the role of the state should be kept to a minimum' (*Agence Europe*, 29 June 1996). A compromise position was reached on this issue in the

March 1997 Telecommunications Council, which increased the number of conditions that could be attached to general authorisations and individual licences; eliminated individual licences from the scope of the approach towards harmonising licensing conditions; extended the granting of individual licences to a larger number of possibilities (for example, voice telephony, mobile telephony and public infrastructure operators); and included numbering resources as grounds for limiting the number of licences (*Reuter European Community Report*, 26 June 1996; *Agence Europe*, 27 June 1996; Kiessling and Blondeel 1998). In line with the decreased influence of individual Member States as a result of qualified majority voting (see chapter five), the directive was adopted even though the German and Greek delegations voted against.[38]

Although users, services providers, MEPs and Commission officials had sufficient resources to form an influential policy network, and were able to gain Council approval of a watered down licensing proposal, they were unable to achieve their preferred policy of a single telecommunications licence. The Member States, in defending their common interest, were able to use their privileged position in the institutional setting to ensure that the national regulators would retain substantial licensing authority.[39] Consistent with the core argument of rational choice institutionalism, that the primary motivation of political actors is to maximise utility through institutional contexts, national government officials used the inter-institutional bargaining stage to increase their ability to implement licensing conditions consistent with national ideologies, institutions and policy-making styles. Moreover, there is no evidence that the process changed their preferences. These rules and procedures also enabled Member State representatives to shift significant licensing authority to the intergovernmental CEPT.

6.3   THE CEPT'S CONTINUED INFLUENCE

The Member States preferred to rely on the intergovernmental institutional framework of the CEPT to the supranational EC for the harmonisation of licensing conditions. A Commission official commented that the Member States are often ambivalent about a single telecommunications market, 'which is why they [were] keen to maintain control over further developments in the intergovernmental context and unwilling to envision an active Community role'.[40] Under the terms of the Licensing Directive discussed above, the CEPT was to play a key role in developing European-level regulations and a one-stop shop for European licences. Nevertheless, this section argues that the CEPT's continued existence and resistance has, instead of aiding licensing harmonisation, further hindered the development of a European-wide licensing regime. This section first examines the CEPT's organisational adaptation, which helps account for some of the difficulties it faced in coordinating European-wide harmonisation. This discussion is then followed by an examination of the CEPT's failure to re-regulate licensing conditions at the European level.

In order to remain a viable, influential institution, the intergovernmental CEPT has had to adapt its original organisational design that was established in 1959 (as discussed in chapter three). New institutionalism explains that the transformation of the CEPT has resulted from an encounter between the CEPT and the environment. Although the most extensive institutional change results from war or other situations of 'punctuated equilibrium' (Krasner 1984), 'the less dramatic version is an ongoing tension among alternative institutional rules – and an ongoing debate or struggle over the matching of institutional principles and actual situations and spheres of activities' (March and Olsen 1989, p. 167). Here, institutional change was prompted by the transformation of the telecommunications sector and the ongoing struggle between competing rule systems.

In particular, the CEPT has been forced to restructure as the Commission has expanded competence and the European telecommunications sector has been transformed. Traditionally, the CEPT was composed of the European PTTs (see chapter three). By the early 1990s, however, the European PTTs were being dismantled; regulatory and operational functions of the telecommunications networks were separated in accordance with the EC's liberalisation efforts (see chapters seven and eight). In response, the CEPT became an organisation of regulators and policy makers.[41] Three separate committees were established: the postal committee (CERP) of postal NRAs; the telecommunications committee (ECTRA) of radiocommunications NRAs, and the radiocommunications committee (ERC) of telecommunications NRAs (see Figure 6.1).

The formation of the ERC, the ECTRA and the CERP was one of the more recent in a series of transformations of the CEPT institutional framework, which began in the 1970s in response to developments at the national and EC levels (see chapters two and three).[42] The focus of this chapter is on the ECTRA. As national PTTs lost their monopoly over telecommunications policy making during the early 1990s, Member State governments sought to create a favourable institutional design for the coordination of transnational telecommunications issues. Intergovernmental cooperation within the ECTRA, and a 1992 Memorandum of Understanding with the EC, ensured that the national regulators would play a prominent role in European-level efforts to harmonise telecommunications regulatory frameworks.

In addition to its intergovernmental, unanimous decision-making process, several other aspects of ECTRA's organisational design and normative

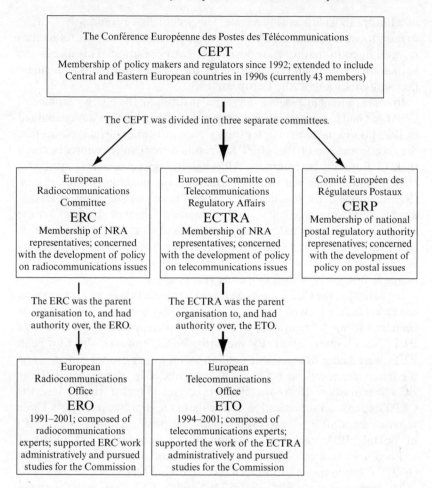

Figure 6.1   CEPT structure 1994–2001

dimension limited its ability to coordinate European harmonisation. First, ECTRA lacked a clearly defined mission. Instead of being proactive towards the rapid changes facing the telecommunications sector, ECTRA generally responded to EC developments. Secondly, ECTRA agreements were non-binding, lacked enforcement mechanisms and did not require members to report compliance. Thus, even when decisions were reached in ECTRA, there was almost no assurance that they would be implemented. Thirdly, ECTRA was less pluralistic and less attentive to the needs of users and service providers than the EC. ECTRA decisions thus typically garnered less support from these key players than EC decisions. Finally, with a membership of 45 European countries, decision making in ECTRA was difficult. Many eastern European members were less liberalised than the EC Member States and added to its conservative nature. These aspects of ECTRA's organisational design make it seem plausible that the Member States, in shifting harmonisation responsibility to ECTRA, sought to limit, rather than enhance, the effectiveness of European harmonisation efforts.

In 1994, ECTRA established the European Telecommunications Office (ETO) in Copenhagen to assist its work administratively and to liaise with the Commission. Formed in response to the initial Commission proposal to coordinate EC licensing regimes (see section 6.2 above),[43] its functions were defined initially to provide the administrative framework for the implementation of the one-stop shopping licensing procedure, and to undertake studies on the harmonisation of licensing and numbering, particularly for the Commission.[44] The fact that the initial functions of the ETO followed closely behind EC-level developments highlights the key role that the Commission played during the 1990s as a policy entrepreneur: the Member States were reacting to developments at the EC level and adapting the intergovernmental CEPT in response to these developments.[45]

Staffed by nine telecommunications experts and funded entirely by NRA contributions and Commission work orders, ETO experts became an important part of the EC-level epistemic community. The institutional design of ECTRA facilitated access to important information, as its members, the NRAs, were often the best sources of information on domestic telecommunications matters. This information enabled the ETO telecommunications experts to produce studies that assisted the Commission in developing the EC's regulatory framework.

At the same time, ETO's parent organisation, ECTRA, was a competing institutional design to the Commission. Since authority over the harmonisation of licensing conditions was shared between ECTRA and the Commission, the intergovernmental nature of the ECTRA frequently put it at odds with the supranational Commission. Caught somewhere in between the two large organisations, the small ETO office often found itself in a precarious position. As an ETO official explained, 'ETO does not always share the views of ECTRA or the Commission. We try to establish our own thinking, which is not always easy if you are a body where the work programme is decided by ECTRA and quite a big amount of our income is paid by the Commission through studies for the Commission.'[46] It is argued below that the CEPT's institutional design and particularly, ECTRA's authority over ETO, further constrained the Commission's effort to harmonise European licensing conditions.

The CEPT as an Inhibiting Factor in the Commission's Harmonisation Efforts

Unlike private sector telecommunications consultants that Commission officials regularly call on for advice, ETO telecommunications experts were hindered in their work by their institutional position within the CEPT and by the national regulators and policy makers who could influence their study results. Since the ETO was dependent on its parent organisation, ECTRA, to approve its contracts with the Commission, Commission officials assert that, although the work order contracts were passed formally between Commission officials and ETO experts, NRA representatives were able to interfere heavily in the definition of the terms of reference of ETO work orders.[47] An ETO official described the Commission-ETO-ECTRA relationship as follows:

> [it] is a strange conglomerate, a mixture of bodies. It's a triangle actually. The Commission pays for the studies, ETO is contracted with the Commission on the basis of a Framework contract, but ECTRA still has a say in these studies because they want to agree to the content of the studies and have a say in what is concluded by ETO.[48]

This convoluted institutional environment, together with the resistance of the NRAs, constrained the development of harmonised licensing conditions. In fact, the NRAs, in several cases, prevented ETO telecommunications experts from completing work orders contracted by the Commission. In so doing, the institutional design of the CEPT enabled the NRAs to further hinder the Commission's harmonisation efforts.

In accordance with the terms of the Licensing Directive (see section 6.2 above), the Commission commissioned 27 studies from ETO between 1996 and 1999. These studies aimed to establish common licensing and numbering regimes in European countries. Nevertheless, ETO finished only one of these studies with complete success; 26 studies failed to comply fully with the terms of reference stipulated in ETO's work orders.[49] For example, several studies were supposed to propose detailed harmonised obligations that could be included in a set of licensing conditions for telecommunications services. However, due to their privileged position within the institutional setting (see Figure 6.1), the national regulatory authorities (through ECTRA) passed a July 1997 Resolution that barred ETO telecommunications experts from including harmonisation proposals in certain reports to the Commission.[50] ETO was unable to complete these work orders and, as a result, the Commission refused to pay the full amount contracted for at least three studies;[51] refused acceptance of at least two other draft final reports; and insisted on the redrafting of several others until they conformed more closely with the terms of reference.[52]

Due largely to the difficulty ETO experts have had in getting the NRAs to provide requested information, 12 studies commissioned by the Commission in 1997 were completed (on average) more than a year past the deadline stipulated in the terms of reference (one final report was delayed more than two and a half years (ETO 1999d)). In nearly every case, Commission officials rejected the initial final reports, accepted the final reports pending editorial corrections, or accepted the final reports with reservations.[53] ETO (1999a) even conceded that the 'harmonisation of licensing regimes or mutual recognition of authorisations seems to be very difficult to implement, due to the reluctance of a majority of national regulators'.

Commission officials maintain that, due to NRA resistance, there were shortcomings with almost all ETO studies. One Commission official commented that there were difficulties with ETO studies regarding 'non-compliance with deadlines, alteration of reports to accommodate ECTRA (but with which the Commission, funding the work order, did not agree), and a general perception of an overall lack of value for money as compared with similar work produced by the private sector'.[54] Another Commission official considered the quality of the ETO reports to be 'unsatisfactory' and complained that 'if a private consultant had produced similar results … [the Commission] would not have commissioned further studies from that consultant'.[55]

The CEPT was also unable to produce the one-stop shopping procedure assigned to it under the Licensing Directive. The Commission (1999, p. 51) cited this as the best example of the failure of the CEPT to produce satisfactory work: 'Existing procedures for co-operation with CEPT/ ECTRA have not worked satisfactorily. Almost without exception, the deliverables supposed to result from this co-operation have not materialised (the most important example of this being the failure to agree on a one-stop shopping procedure in the field of licensing).' Market players also complained to the Commission of their disappointment in the CEPT's failure to develop the one-stop shopping procedure (Commission 1999b).

In sum, since the Member States ensured that the CEPT was given a key role in European licensing harmonisation (see section 6.2 above), national regulators retained power and were able to hinder the Commission's harmonisation efforts. During the mid- to late-1990s, therefore, the NRAs used their institutional position within the CEPT to hinder the harmonisation of national licensing conditions at the European level. As a result, and as the discussion below indicates, the Member States continued to maintain diverse national licensing regimes.

6.4   CONTINUED DIFFERENCES IN LICENSING CONDITIONS

Historical institutionalism recognises the value of analysing the implementation of policies, because implementation often leads to a new round of policy making. Armstrong and Bulmer (1998, p. 57), for example, explain that historical institutionalism 'rejects the notion that politics can be separated from public administration, for putting of policy into practice is an essential part of the whole: the experience of administration may start a new cycle of policy development'. It is particularly important to evaluate policy implementation in examining EC policy making. Since the EC is without its own administrative agencies at the national level, the Member States, through their NRAs, are left in control of the implementation and execution of EC policies.

The joint-decision trap explains that this institutional arrangement reduces the effectiveness of implementation and results in the lowest common denominator of harmonisation (Scharpf 1988). The approach helps to account for the ability of the Member States to implement the Licensing Directive in ways that are consistent with national ideologies, institutions and policy-making styles. Richardson (1996b, p. 285) similarly explains that the EU is attempting 'to weld fifteen different "regulatory styles" into a European system of regulation. It is not surprising that the regulatory patchwork that results does not "fit" the regulatory and implementation systems of each of the member states.' As a result, following the deadline for its implementation, the Member States continued to maintain a patchwork of distinctive national licensing regimes.[56] Because of this institutional environment, and the resistance of national policy makers and regulators discussed in previous sections, Member States were able to retain disparate licensing conditions despite the harmonisation intent of the Licensing Directive.

Analyses of the implementation of the Licensing Directive find that Member States retained a significant amount of divergence in the conditions for licensing telecommunications operators (see ETO 1999a).[57] The Commission (2000, p. 2) explained that 'studies conducted by the European Telecommunications Office on mandates by the European Commission have made clear that in the Community today there is no harmonised approach to authorising market entry for communications service providers but a patchwork of fifteen national regimes which are widely divergent in their basic approach and specific detail'. The Commission's November 1999 Implementation report also found that the goal of the Licensing Directive had not been achieved and that there was little harmonisation of national licensing regimes (Commission 1999a).

In the more liberal Member States, such as Denmark, the Netherlands, Finland and Sweden, the licensing regimes are extremely light and operators are able to enter the market with little inconvenience. In the UK, operators are also able to enter the market easily, although the licensing conditions are more extensive. ETO (1999a, p. 88) explains, 'the UK is an exception. Notwithstanding high segmentation, high use of individual licenses and high quantity of information to be provided, the UK remains a country where the degree of difficulty in obtaining an authorisation is quite low.' In line with the British policy style examined in chapter two, and the emphasis placed on increasing competition, the UK's peculiar approach to licensing stems from its desire to prevent the powerful BT behaving anti-competitively.[58]

On the other hand, the less liberalised Member States generally maintained heavy licensing regimes into the late 1990s. In many Member States, 'individual licenses are the rule and in some cases a government minister is required to sign every license. ... onerous conditions going far beyond the letter and spirit of the directives are laid down in the licenses themselves, and in some cases are entirely confidential as between the issuing authority and the operator concerned' (Commission 1999a). Belgium, for example, required potential operators to submit comprehensive business plans covering several years into the future. 'The licensing procedures are burdensome and imply a great amount of work for the applicants' (ETO 1999a, p. 88).

France also maintained licensing conditions that went beyond those established in the Licensing Directive. This can be explained by reference to its national policy style examined in chapter two. ETO (1999a, pp. 86–87) explains that:

> the conditions attached to the authorisations are burdensome and in some cases can constitute market entry barriers. The fragmentation of the licensing scheme into many different definitions of networks and services with their related licensing conditions and procedures seems to be the result of the traditional economic and political state-control over telecommunications in the country.

In this way, reform of the telecommunications sector in France continued to be stickier than in the UK. The enduring influence of *dirigisme* on French political and institutional actors led the national regulator to retain a great degree of control over market access. Meanwhile, the French preoccupation with telecommunications service as a *service publique* led French policy makers to implement a universal service fund, which required licence fees significantly higher than costs.

Reform in Germany also continued to lag behind the UK. Despite the independence granted to the NRA in Germany, it was not set up until the beginning of 1998 and, even then, the German government maintained a degree of influence over its decisions (see chapter two). The processing of licences generally exceeded the six weeks mandated in the Licensing Directive (Commission 1999a). Germany, along with France, maintained 'unreasonable and unjustifiably high license fees' (ETO 1999a, p. 103; see also Commission 1999a).[59] Potential new entrants considered this to be a condition that could impede entrance into the market.

Institutional path dependencies help account for the potential entry barriers to the telecommunications market that new entrants continued to face. '(T)he fragmentation of the licensing scheme ... is in most cases the result of traditional and historical state-intervention and state-control over the telecommunications sector, characterised by activities of "prominent general interest"' (ETO 1999a, p. 109).[60] European capacity for positive integration in this sector, as in other sectors, is therefore constrained by diverse and historically contingent national institutions, values and practices.[61] Although these national structures are continually evolving in response to exogenous pressures, this change has proven to be path dependent.

Although a lack of harmonisation may have negative economic consequences,[62] this section does not argue that the harmonisation of the European telecommunications sector has been unsuccessful. On the contrary, most of the ONP regulatory framework (including the ONP Voice Telephony and Framework Directives examined in chapters four and five) was very successful and was largely implemented on schedule.[63] This chapter has, however, pointed to an area of policy where the Member States relied on their advantageous position in the institutional setting to defend a common interest and impede the shift of regulatory authority to the European level. It demonstrates that the Commission cannot force Member States to accept a major decision for positive integration and that there are limits to the Commission's capacity to bridge the gaps between Member States' positions. As Scharpf (1999, p. 191) concludes, 'there are definite limits to the possibilities of a "deliberative" redefinition of government preferences'.

Due to their institutional authority for positive integration, the Member States were able to prevent the establishment of an ERA and the implementation of a coherent EC-level licensing regime. Thus, notwithstanding the Commission's widespread success as a policy entrepreneur, it has indeed faced limits in its ability to Europeanise the sector.

6.5   CONCLUSION

This chapter found that institutional settings and resources affect the nature of policy networks. Although Commission officials, MEPs, users and service providers favoured the establishment of an ERA, these actors lacked sufficient resources to form an influential policy network in pursuit of this objective. As a result, they were unable to get the issue actively considered on the EC's agenda. On the other hand, national regulators and policy makers had a large degree of policy-making authority. They formed a powerful intergovernmental policy community and have been able to prevent formal negotiations over an ERA from occurring at the EC level.

In a different institutional setting provided for under the co-decision procedure, however, Commission officials, MEPs, users and service providers were able to form an influential policy network in the pursuit of licensing harmonisation. As a result, the policy network was successful in gaining the Council's approval of a Licensing Directive in 1997. Thus, the characteristics of these networks and the ability of key actors to control or determine outcomes were both significantly affected by the institutional environment in which policy is made. Nevertheless, notwithstanding the network's success, it still lacked sufficient resources to achieve its favoured policy of a single telecommunications licence or the mutual recognition of licences. The existence, organisational design and resistance of the intergovernmental CEPT framework further hindered the development of a coherent set of harmonised licensing conditions. As a result, even following implementation of the Licensing Directive in 1998, Member States maintained widely divergent national approaches to the licensing of telecommunications operators.

The above case study demonstrates the difficulties faced by European policy makers in attempting to achieve positive integration, thus confirming Scharpf's (1999, p. 71) assertion that the 'problem-solving capacity of positive integration is limited by the need to achieve action consensus among a wide range of divergent national and group interests'. Scharpf's joint-decision trap (1988) is particularly useful for highlighting the power relations present in EC institutions. Due to the requirement for unanimous Member State agreement, national government representatives have greater control than other actors in policy development and in creating and modifying institutions. As a result, the Member States have been able to resist the establishment of an ERA and a coherent EC-level licensing regime.

NOTES

1. Commission (2000) Draft Directive on the authorisation of electronic communications networks and services, COM(2000)386, 12 July, pp. 4–5.
2. Keleman (2002) describes the establishment of agencies at the European level as 'one of the most notable recent developments in EU regulatory policy'. Indeed, since 1990, there have been eight European agencies established in matters of economic or social regulation, including environmental protection, trademark and design registration and pharmaceutical regulation (*Ibid.*).

3. See chapter two for a discussion of OFTEL's powers.
4. Interview, former ECTRA official, Brussels, June 1999; Interview, INTUG representative, telephone, November 1999; Analysys Consultants (1989), p. 68; and *Communications Week International*, 24 June 1991.
5. The European Space Agency is an intergovernmental organisation that includes most EC Member States. It has maintained a programme in telecommunications since 1970.
6. Commission (1994) 'Communication on satellite communications: the provision of – and access to – space segment capacity', COM(94) 210 final, 10 June 1994.
7. Quoted in *ibid.*, italics in original.
8. 'Communication on the Consultation on the Green Paper on Mobile and Personal Communications', COM (94) 492 final, 23 November 1994.
9. Interview, MEP Ferdinand Herman, Brussels, May 1999; Interview, MEP Mel Read, telephone, July 2000. In a 2002 letter to the *Financial Times* (8 January), MEP Renato Brunetta wrote of the need to establish an ERA.
10. The EP proposed amendments to both the 1996 ONP Interconnection and 1997 Licensing Directives.
11. Interview, DG Information Society (DG XIII) official, Brussels, March 1999; Interview, DG Competition (DG IV) official, Brussels, May 1999.
12. Michel Carpentier speech to the *Financial Times* World Telecommunications Conference, quoted in *The Financial Times*, 9 December 1993.
13. Bangemann Report 'Europe and the Global Information Society: Recommendations to the European Council' 26 May 1994.
14. Commission (1994) 'The Information Society in Europe: a first assessment since Corfu' (The 'Essen Conclusions'), 9–10 December.
15. Commissioner van Miert keynote address to the IIC Telecommunications Forum 'Preparing for 1998 and Beyond', 15 July 1996, quoted in EC Press Release (96–198).
16. See 'Next challenges facing the 1998 European telecommunications regulatory framework', speech at the EU Competition Workshop, Florence, 14 November 1998, where Director-General Verrue indicated that building on a network of NRAs, 'not always substituting certain functions but offering complementary alternative service, could be a way to proceed towards a European authority'.
17. Article 308 can be used to achieve one of the objectives of the Treaty in the event that authority to achieve that action is not provided for elsewhere in the Treaty. For a detailed analysis of the Treaty considerations for the establishment of an ERA, see NERA (1997).
18. Article 5 of the EC Treaty (formerly Article 3b) allows EC action 'only if and insofar as the objectives of the proposed action cannot be sufficiently achieved by the Member States and can therefore, by reason of the scale or effects of the proposed actions, be better achieved by the Community'.
19. Worthy and Kariyawasam (1998, p. 7) assert, 'It is clear that if an ERA is to be formed and assume a licensing role … [the] present goal posts on subsidiarity need to be moved'.
20. 'Unraveling Europe's telephones: The case for a regulator', *The Economist*, vol. 333, 3 December 1994, pp. 18–20.
21. For a discussion of subsidiarity in the context of European integration, see Mazey (2001).
22. COM(92)254 'Proposal for a Council Directive on the mutual recognition of licenses and other national authorisation to operate telecoms services, including the establishment of a Single Community Telecoms License and the setting up of a Community Telecoms Committee' (25 September 1992).
23. EC Press Release, 'Proposals for action in the telecommunications services sector', IP/585.
24. *Ibid.*
25. This is the same strategy that the Commission pursued with the national PTTs in forming the SOGT and GAP committees in 1983 (see chapter three).
26. 'Memorandum Francais sur l'Europe des telecommunications' presented to the Telecommunications Council, 10 May 1993. Translated as '*The principle of subsidiarity*

must be in particular the base of a harmonious distribution of roles between Community and Member States. The application of the Community disposition must be thus confided with the authorities of national regulations, notably for the management of regulatory problems (licenses, numbering, frequencies, taxes (tariffs); the necessity of cooperation for resolving the questions of pan-European dimension ought to support such a possibility over the CEPT.'

27. Interview, Commission official, Brussels, May 1999.
28. COM(93)652 'Proposal for a Directive by European Parliament and Council on a policy for the mutual recognition of licenses and other national authorisations for the provision of satellite network services and/or satellite communications services', 4 January 1994.
29. Commission (1994) 'The Information Society in Europe: a first assessment since Corfu' (The 'Essen Conclusions'), 9–10 December, italics in original.
30. Commission of the European Communities, Amended Proposal for a Licensing Directive, 'Proposal for a European Parliament and Council Directive on a common framework for general authorizations and individual licenses in the field of telecommunications services', COM (95)545 final (96/C 90/05) – 95/0282(COD) Submitted by the Commission on 30 January 1996.
31. EC Press Release (IP/95/1234) 'Commission adopts two proposals completing the regulatory framework for a liberalised telecommunications market', 14 November 1995.
32. EC Press Release (MEMO/95/158) 'Telecommunications liberalisation: state of play', 27 November 1995.
33. COM(96) 342 final, July 31, 1996, OJ No C 90, 27.3.1996.
34. Council of Ministers Common Position on Licensing Directive, Common Position (EC) No 7/97 adopted on 9 December 1996 (97/C 41/04).
35. Council of Ministers Common Position on Licensing Directive, Press Release (97/53), 6 March 1997.
36. EP Session News Press Release, 18 February 1997; and European Report, 22 February 1997.
37. Council of Ministers Common Position on Licensing Directive, Common Position (EC) No 7/97 adopted on 9 December 1996 (97/C 41/04).
38. Council of Ministers Press Release (97/53), 6 March 1997; Directive of the European Parliament and of the Council on a common framework for general authorisations and individual licenses in the field of telecommunications services COM 97(17), 10 April 1997, OJ L 117, 17 May 1997, p. 15.
39. For a more detailed examination of the potential scope of NRA freedom of action under the licensing directive, see Kiessling and Blondeel (1998) or Xavier (1998).
40. Interview, DG Information Society (DG XIII) official, Brussels, March 1998.
41. As chapter four explained, ETNO took over as the political representative of the national telecommunications operators (former PTTs).
42. As discussed further below, the CEPT underwent a further institutional transformation in 2001, which resulted in the merger of the ERC and ECTRA.
43. Commission Proposal for a Licensing Directive, European Commission, COM (92) 254 final – SYN 438 (15 July 1992).
44. ECTRA (93) 69 final 'Memorandum of Understanding on the establishment of the ETO among ECTRA members'.
45. As discussed further below, the CEPT underwent a further institutional transformation in 2001, which resulted in the merger of the ETO and the ERO.
46. Interview, ETO official, telephone, July 2000.
47. Interview, DG Information Society (DG XIII) official, Brussels, June 1999.
48. Interview, ETO official, telephone, July 2000.
49. Interview, DG Information Society (DG XIII) official, Brussels, June 1999. The only work order that complied fully with the terms of reference stipulated in the work order is ETO (1996) 'Fixed packet or circuit switched data services offered to the public'. This work order, however, was submitted in the context of the draft mutual recognition of

licences directive. Since the Council did not accept the mutual recognition approach, the results of this work order were not translated into legal obligations.

50. Interview, DG Information Society (DG XIII) official, Brussels, June 1999; Interview, ETO official interview, telephone, July 2000; the ETO official pointed out that the decision was specific to three studies, although some ECTRA members 'used this opportunity to create a crisis in a more generalised form, that there was no support for harmonisation measures anymore'.

51. The three studies are ETO (1998) 'Final report on regulating operators with significant market power', ETO (1998) 'Final report on harmonising essential requirements' and ETO (1998) 'Final report on consumer protection'.

52. Interview, DG Information Society (DG XIII) official, Brussels, June 2000.

53. Interview, DG Information Society (DG XIII) official, Brussels, June 2000.

54. Interview, DG Information Society (DG XIII) official, Brussels, June 2000.

55. Interview, DG Information Society (DG XIII) official, Brussels, June 2000; Interview, DG Information Society official, telephone, September 2004.

56. For a discussion of the difficulties in developing an EU regulatory policy that accommodates diversity, see Héritier (1996).

57. For a discussion of the lack of harmonisation in interconnection approaches, see Osborne (2000). With regard to interconnection, ECTRA members prevented ETO from fulfilling the terms of a CEC work order on a study (ETO 1998), which would have included a harmonised approach to calculating interconnection rates. Page six of the study explains that 'ECTRA [does] not see any need for further harmonisation with regard to any of the obligations on operators with significant market power'.

58. Interview, OFTEL representative, London, May 2000; Interview, DTI representative, London, May 2000.

59. German interconnection arrangements were also criticised as a potential barrier to market entry.

60. See chapter two for a discussion of traditional state intervention in the telecommunications sectors of the UK, France and Germany.

61. For examples of the limits to effective Europeanisation in other sectoral studies, see Levy (1999); Thatcher (1997b); and Rhodes (2000).

62. For example, the ETO's examination of the Member State licensing regimes concluded that these 'national peculiarities may threaten the potential benefits of the single market for telecommunications' (ETO 1999a, p. 107). Similarly, Osborne (2000) concluded that the differences between European interconnection approaches would lead to lower competitive intensity, less innovation and higher prices, and would harm consumers and the competitiveness of industry.

63. See the Commission's 3rd, 4th, 5th, 6th, 7th and 8th 'Implementation Report on the 1998 Regulatory Package'. The 8th Implementation Report concludes that the regulation put in place at the national level was very substantially compliant with the EU framework. COM(2002)695 final, 3.12.2002.

PART III

The liberalisation of European telecommunications

7. Applying EU competition law to the telecommunications sector

Competition policy is necessarily evolutionary in nature and must respond to the changing European and world economic, social and political climate if it is to remain relevant and vital ... This does not mean there are no basic rules or principles. On the contrary, these are set out clearly in the Treaty ... which have been agreed unanimously by Member States. Rather these rules must be applied in a realistic and pragmatic way.[1] *Competition Commissioner Karel van Miert, May 1993.*

INTRODUCTION

Part III of this book analyses the authority of national governments and institutional actors under supranational European law (Weiler 1982) or negative integration (Scharpf 1999, ch. 2). Following the two-pronged strategy put forth in the 1987 Green Paper on telecommunications (see chapter three), the Commission set out to liberalise the EC telecommunications sector. This chapter, in particular, examines the initial efforts, from 1988–1993, of DG Competition (formerly DG IV) to apply the EC Treaty rules on competition policy to the telecommunications sector. In line with Fritz Sharpf's (1999, p. 49) conclusions, it confirms that 'the institutional capacity for negative integration is stronger than the capacity for positive integration'. Therefore, in contrast to the previous chapter, which found that the Commission lacked sufficient institutional authority to push through some of its initiatives, this chapter argues that the prevailing rules and procedures for competition policy facilitate Commission efforts to achieve its desired policy line.

This chapter is divided into three substantive sections. The first section gives a brief introduction to the institutional setting provided by the EC Treaty rules on competition. The second section demonstrates that this institutional context provided the Commission with substantial formal agenda-setting authority; DG Competition officials were able to push through a 1988 directive liberalising terminal equipment and a 1990 directive liberalising value-added services over the resistance of several Member States, including France and Germany. The ECJ played a key role during this period by confirming Commission authority to issue these directives unilaterally. Demonstrating the importance of rules and changes in rules, the third section argues that the institutional balance of power nevertheless shifted towards the Member States in 1992 as the EC institutions experienced a normative shift towards respecting subsidiarity. Highlighting the ongoing evolution of rules reflecting inter-institutional battles, the Member States and affected interests subsequently played a more central role in the formulation of a Commission proposal to liberalise voice telephony services.

7.1   THE COMMISSION'S COMPETITION AUTHORITY

Commission efforts to coordinate negative integration, or the elimination of trade barriers between Member States, benefit tremendously from supranational EC law. Negative integration has been a fundamental objective of the Community since its inception in 1956. The EC Treaty established that a basic Community activity should be to put in place a system to safeguard against distortions of competition within the Common Market. The EC Treaty rules on competition (Articles 81 to 89 of the EC Treaty, formerly Articles 85 to 94) provide an institutional context under which the Commission maintains extensive authority to establish this system. The focus of this chapter is on the authority of the Commission under Article 86 of the EC Treaty (formerly Article 90). Although Article 295 of the EC Treaty (formerly Article 222) ensures that the Member States may continue to govern their national system of property ownership (that is, the Commission cannot therefore require the privatisation of national TOs), Article 86 subjects public and private firms granted special or exclusive rights (for example, monopoly rights) to the EC competition rules.

In contrast to the proposals analysed in Part II of this book, proposals drawn up under Article 86 are not subject to the ONP consultation mechanisms set up by the Commission (examined in chapter four), nor are they required to undergo the institutional bargaining stage set up by the EC Treaty (examined in chapter five). Instead, an extraordinary clause in Article 86(3) empowers the Commission to issue decisions or directives unilaterally. As Craig and De Búrca (1995, p. 121) note, the Commission can 'make legislation without any intervention from the other institutions'. These Commission directives are therefore binding on the Member States even though they do not require the formal approval of national governments. Part III of this book finds that this institutional context was crucial in giving the Commission power and facilitating the achievement of its desired policy line: the liberalisation of the EC telecommunications sector.

The motivation of the Member States to transfer this extensive authority to the Commission set out under Article 86(3) is consistent with the principal-agent model of delegation (Pollack 1997) and liberal intergovernmentalism (Moravcsik 1994). They sought to minimise the costs and maximise the gains of mutual cooperation by delegating authority to a supranational agent, the Commission. In particular, Germany, Belgium, the Netherlands and Luxembourg insisted on including Article 86 in the Treaty of Rome in order to prevent trade distortion from the comprehensive French and Italian public monopolies. Joint gains were perceived to be large in controlling the French and Italian public monopolies, whereas efforts to secure compliance by their national governments would probably have been ineffective. Enabling the Commission to act as a supranational agent under Article 86(3) thus enhanced the credibility of this commitment. The political compromise in drafting the article is apparent in 86(1), however, which acknowledges the right of Member States to maintain special and exclusive rights that do not contravene the Treaty, and 86(2), which prevents the Treaty from obstructing the role of utilities in exercising those rights.

In line with historical institutionalism, however, the application of Article 86 to the telecommunications sector is an unanticipated consequence (Pierson 1996) of the drafting of the Treaty of Rome. Indeed, none of the Member States expected that Article 86 would apply to the telecommunications sector or any other public service sector.

> Historically, there is no question that neither the governments negotiating the Treaties of Rome nor the parliaments ratifying them had any intention to use European competition law to challenge the existence of *service public* functions in the member states of the Community. In fact, Art. (86) II TEC contains language which suggests exactly the opposite. (Scharpf 1999, p. 63)

The ECJ ruling in the 1974 Sacchi case was subsequently interpreted by EC lawyers as confirming that utilities were national matters and beyond the scope of the internal market.[2] Because of its controversial nature, the Commission did not use Article 86 extensively until the late 1980s.[3] This chapter finds that almost all Member State governments, including the UK, France and Germany, initially opposed the application of Article 86 to the telecommunications sector. Despite this opposition, the institutional opportunity structure nonetheless enabled the Commission to achieve its desired policy line.

In addition, the paradigms, values and beliefs embedded within the Commission, such as its pro-integration mission and the pro-competition mission of DG Competition (formerly DG IV), influenced the aims and motives of Commission officials (see March and Olsen 1984, 1989). Indeed, the decision to apply the treaty rules on competition to the telecommunications sector was closely tied to the institutional values and norms of the EC and, particularly, DG Competition. DG Competition officials have a common belief in the merits of competition and feel that they are on 'the moral high-ground, endowed with a mission to establish norms and to encourage working practices that promote competition' (Cini 1997, p. 86). DG Competition officials viewed the telecommunications sector as analogous to any other industrial sector. Therefore, they felt that it should be subject to the Treaty's competition rules. Normative considerations thus help account for the willingness of DG Competition officials to employ Article 86 to liberalise EC telecommunications. In addition, in keeping with the bureaucratic models of politics discussed in previous chapters, DG Competition also wishes to increase its power.

Although DG Competition had been granted extensive competition powers under the Treaty of Rome, 'by the late 1970s it had the appearance of a policy that had become sidelined ... advocates of a strong competition policy seemed to be swimming against an ideological tide ... [DG Competition] staff consistently found themselves at odds with the nationally-orientated interventionist spirit of the times' (Cini 1997, p. 76).[4] During the 1980s, however, the revitalisation of the aim to complete the internal market (see chapter three) enhanced DG Competition's competition authority and helped to make Article 86, which had previously been a latent institution (Thelen and Steinmo 1992), become salient. In exploiting the treaty rules on competition, DG Competition was able to bolster its own authority and strengthen its role in the polity.

During the 1980s, DG Competition also recruited heavily from the national competition authorities, which increased the perceived legitimacy of its actions due to the recognised expertise of its staff (Cini 1997, p. 77). As Radaelli (1995 p. 179) has pointed out, 'knowledge can change the perceived characteristics of the games'. In this case, knowledge-based policy development enhanced DG Competition's credibility and thus increased its ability to coordinate liberalisation of the telecommunications sector. Recruiting from national competition authorities may also have facilitated EC policy making because these national representatives were, to some extent, able to smooth objections and pave the way for the EC's efforts.[5]

Following the establishment of the legal norm that competition rules are fully applicable to the telecommunications sector, as a result of the 1985 British Telecom case (see chapter three), a policy window opened for DG Competition. It thus proceeded to intervene in Germany, Belgium, Italy, the Netherlands and Denmark over illegal efforts to extend exclusive rights to new forms of terminal equipment, including cordless telephones and modems (see Commission 1987, pp. 124–129; Schmidt 1998, p. 173). In this way, historical institutionalism highlights the impact that legal norms and the 'accumulation of jurisprudence' have on institutional norms (Armstrong and Bulmer 1998, p. 52), which helps account for DG Competition's determination to ensure that competition was introduced into telecommunications.

DG Competition officials felt that the only way they could achieve the liberalisation of telecommunications by 1992 (the deadline for completion of the internal market) was through the use of Article 86.[6] The prevailing rules and procedures thus affected the perceptions and preferences of political actors as DG Competition officials felt obliged to utilise their extensive competition authority to facilitate liberalisation. Competition Commissioner Brittan asserted, 'Article [86(3)] imposes a duty on the Commission to police the application of this provision. The Commission has no discretion in the matter.'[7]

7.2   THE LIBERALISATION OF TERMINAL EQUIPMENT

As chapter three explained, the liberalisation strategy was part of the two-pronged Commission strategy outlined in a 1987 Green Paper. The dual strategy to liberalise and harmonise EC telecommunications gained the support of a large network of influential actors through a pluralistic consultation process. In pursuing the liberalisation of terminal equipment, the Commission took advantage of a trend that was occurring in all Member States. 'The trend in all Member States is towards progressive full opening of the terminal market to competition' (Commission 1987, p. 61). As explained in chapter two, the UK had liberalised fully in the early 1980s, while both France and Germany had largely liberalised their terminal equipment markets in the late 1980s. Table 7.1 demonstrates that, in 1987, the Member States allowed private suppliers to provide most types of terminal equipment (with the notable exception of the first telephone set).

Although every Member State was moving towards the full liberalisation of terminal equipment, Table 7.1 demonstrates that significant differences in national regulatory conditions remained. DG Competition felt that these differences could have resulted in delays if the Commission had attempted to achieve full liberalisation through the normal legislative procedures (as examined in Part II of this book). The Commission claimed that the institutional opportunity structure provided under Article 86 was therefore justified, because 'any alternative would have been too time-consuming,

Table 7.1 Member State terminal equipment regulatory conditions in 1987

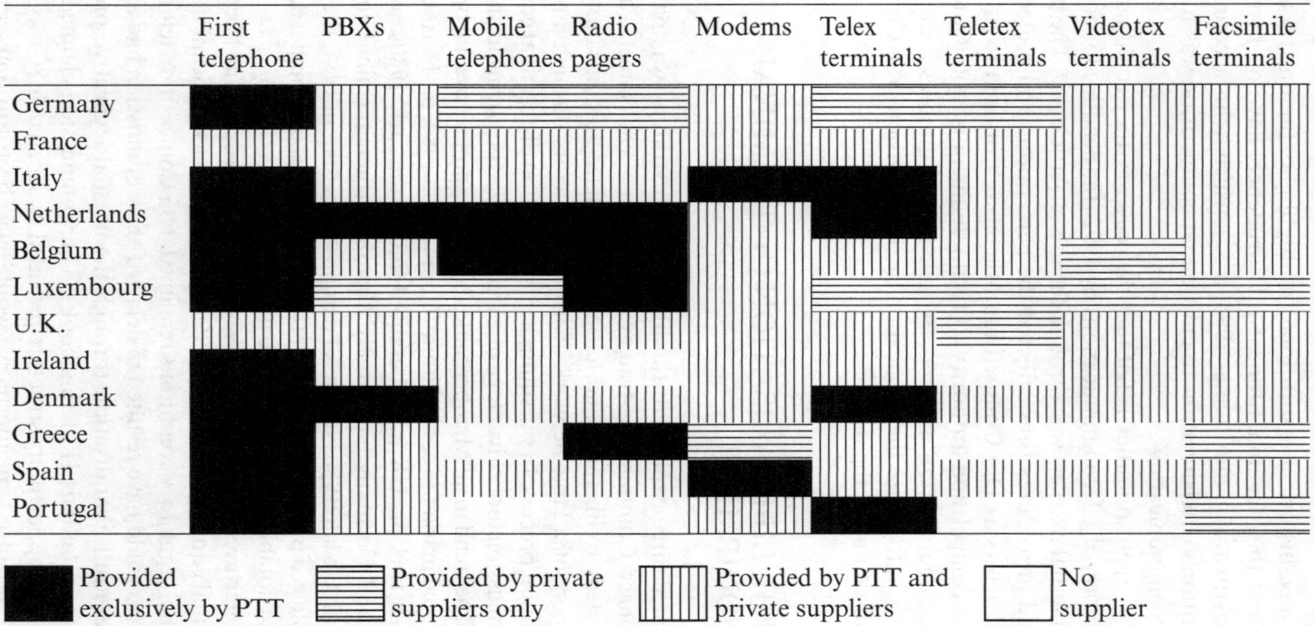

First telephone
PBXs
Mobile telephones
Radio pagers
Modems
Telex terminals
Teletex terminals
Videotex terminals
Facsimile terminals
Germany
France
Italy
Netherlands
Belgium
Luxembourg
U.K.
Ireland
Denmark
Greece
Spain
Portugal
Provided exclusively by PTT
Provided by private suppliers only
Provided by PTT and private suppliers
No supplier

Source: Figure 8 of Commission (1987). The Commission notes that tables such as this are 'inevitably a simplification'.

given the imminence of 1992: direct action was "necessary"' (Cawson *et al.* 1990, p. 195).

DG Competition therefore drafted an Article 86 proposal for the introduction of full competition to the terminal equipment market (the Terminal Equipment Directive) in early 1988. The Member States were initially caught off guard by the Commission's use of this new legislative instrument and ultimately opposed it. Following a Commission presentation of the draft proposal to an informal Council meeting in Berlin in April 1988, a DG Competition official noted that the Member State representatives largely agreed with the contents of the proposal, but that they looked forward to examining it further in a Council working group. The national officials were 'shocked' when a DG Competition official told them that, 'No, you won't be discussing it again. Next you will read it in the Official Journal!'[8] Although the substance of the proposal was not contentious among the Member States, the UK, Germany and France (with the support of every other Member State except the Netherlands), insisted that it had to be proposed to the Council officially.[9]

In utilising Article 86, the draft directive was not subjected to extensive public consultation in the policy formulation stage (as examined in chapter four) nor did it proceed through the Community institutions under the complicated rules and procedures of the inter-institutional bargaining stage (as examined in chapter five). Instead, Article 86 provided DG Competition with the authority to formulate, adopt and publish the Terminal Equipment Directive unilaterally and exceptionally quickly in May 1988.[10] In the same way that disputes over institutional authority are a salient feature of the inter-institutional bargaining stage (see chapter five), the contentious nature of the Article 86 legislative procedure sparked a legal debate between the Commission and the Member States. In June 1988, France, with the support of Belgium, Italy, Germany and Greece, filed a motion in the European Court of Justice (ECJ) contesting the institutional authority of the Commission to issue the Terminal Equipment Directive.

Although UK officials had initially expressed opposition to the choice of legislative procedure, the UK government did not support referral to the ECJ. Even though the British government has historically sought to limit the transfer of competences to the EC, in this instance British interests were in line with Commission objectives. As with the British preference for a Commission-dominated regulatory committee to oversee implementation of the ONP legislation (see chapters four and five), British national interests lay with increased Commission authority to pursue the liberalisation agenda. A British COREPER official explained that Article 86 'kickstarted the process ... [and] was a necessary first step towards liberalisation, but not something we would instinctively welcome'.[11] The support of a powerful Member State, the UK, nonetheless facilitated DG Competition's efforts.

7.3   THE LIBERALISATION OF VALUE-ADDED SERVICES

While the Terminal Equipment Directive was being challenged in the ECJ, DG Competition officials began working on a second Article 86 draft directive (the Services directive) aimed at removing all special or exclusive rights in the supply of value-added telecommunications services.[12] In order to ensure effective implementation of the directive, the proposal also required Member States to ensure the separation of operational and regulatory activities. Although the directive was set for adoption in April 1989, the Commission sought to accommodate the Member States by delaying adoption until the Council completed work on the complementary ONP Framework Directive (examined in chapters three and four). Competition Commissioner Brittan nonetheless warned that the Commission was prepared to adopt the Services Directive unilaterally if the Member States continued to delay. Brittan commented that he hoped:

> that EC industry ministers will approve a common position on ONP by 1 April 1990. If that target is met, the Commission will time the entry into force of the Services Directive with the formal approval of the ONP Directive ... If the Council does not follow this timetable ... the Services Directive will nevertheless become effective April 1 [1990].[13]

In the 'extraordinary' Telecommunications Council of 7 December 1989, the Commission agreed to modify certain aspects of the Services Directive in order to gain Council approval on the ONP Framework Directive.[14] Highlighting the different power relations in the two institutional settings examined in this book, the Member States used their increased institutional authority under the EC Treaty rules for positive integration (as examined in Part II of this book) to gain concessions in an area of negative integration. The Commission likewise used its increased authority under negative integration to pressure the Member States into agreeing more quickly on a proposal for positive integration.

Demonstrating once again that Member State representatives are rational actors seeking to defend national interests at the EC level, the UK, Germany, Denmark and the Netherlands agreed with the Commission's effort to liberalise all telecommunications services except voice telephony. As chapter two explained, the British government had liberalised telecommunications services in the early 1980s, while Germany had achieved significant liberalisation in value-added services by the middle of 1989. On the other hand, France, which had not liberalised value-added services, along with the seven other Member States, also sought to exclude data transmission services from the scope of the directive (Wheeler 1992, p. 85). Since the latter group of countries maintained a majority, Commission officials modified the proposal in order to allow the PTTs to maintain monopolies over data transmission until 1992.

Although the Commission had the institutional authority to adopt the directive unilaterally, DG Competition officials nevertheless sought to appease the Member States. Due to uncertainty over the results of the pending ECJ case on the Terminal Equipment Directive, Commission officials did not want to provoke a large number of PTT Ministers further. Significant national government opposition to Commission plans could, potentially, lead to the application of sanctions against the Commission, such as treaty revisions reducing Commission authority or widespread non-compliance with Commission initiatives. DG Competition officials understood that, despite their own interpretation of the EC Treaty rules on competition, it was ultimately the Member States that give them their power and that this power could be reduced or even revoked. Politics, therefore, are important in the exercise of legal authority.

As a result, the Commission made further concessions to Member State demands. In line with the French *dirigisme* examined in chapter two, France gained the right to impose public service obligations (such as coverage requirements and quality on data service providers) if it was in the general economic interest. DG Competition nonetheless maintained the authority to scrutinize such obligations to ensure that they were objective, nondiscriminatory and proportionate. The less developed Member States also gained a derogation, which allowed them until 1996 to comply with the terms of the Directive.

These compromises enabled the Commission to gain the support of almost all Member States on the liberalisation intent of the proposal. The UK and Germany introduced a joint statement to the June 1990 Council meeting, which confirmed that a large majority of Member States agreed on the contents of the Services Directive (*Agence Europe*, 30 June 1990). On the other hand, most of the Member States once again objected to the Commission's unilateral adoption of the directive. 'Telecommunications changes', the Council argued, are 'too important a matter for short cuts' and should therefore be processed through both the European Parliament and Council and subjected to full debate and necessary amendment.[15] Following Commission adoption of the Services Directive in June 1990,[16] Spain, with the support of France, Belgium and Italy, filed a complaint to the ECJ contesting the Commission's authority to issue it unilaterally.

ECJ Activism

Many analysts have identified the ECJ as a significant supranational policy actor (see Burley and Mattli 1993; Weiler 1982; Wincott 1995; Dehousse 1998). Dehousse (1998, p. 66–67) claims that 'it is difficult to overemphasize the important contribution the ECJ has made to the integration process'. The ECJ has established 'the authority of Community law and conferred upon it a degree of effectiveness unparalleled in international law'. Burley and Mattli (1993) have put forth a supranationalist or neofunctionalist account of the ECJ in the process of European integration. They assert (p. 74) that 'the Court does have the power to pursue its own agenda, and that the personal incentives in the judicial and legal community, as well as the structural logic of law, favor integration'.

In using its supranational authority to define better the institutional setting under the treaty competition rules, the ECJ has played a key role in the process of liberalising the European telecommunications sector. Beginning with the 1985 British Telecom case (discussed in chapter three), the ECJ has consistently ruled that the Community's competition rules are fully applicable to the telecommunications sector. The fact that the ECJ has consistently interpreted these rules in line with the Commission supports the argument that the ECJ has an institutional bias for integration.

Support for this argument is also found in the Advocate General's opinion on the Terminal Equipment Directive. This independent advice, issued more than a year prior to the ECJ ruling, is supposed to represent an impartial interpretation of the EC Treaty. Craig and DeBúrca (1995, p. 73) note that the opinion of the Advocate General 'is not binding on the court, but it is generally very influential, and the opinion is indeed followed by the Court in the great majority of cases'. In February 1990, Advocate General Tesauro determined that Commission use of Article 86 for the Terminal Equipment Directive was illegal because it 'modifies the very conditions of the presence of the State in any given sector, appears to affect the institutional balances and cannot therefore be considered as legal by the Court'.[17] The significant institutional implications of the legislative procedure thus led the Advocate General to agree with those Member States which had filed the complaint (including France and Germany), and to recommend that the Terminal Equipment Directive be nullified.

However, as with the unclear EC Treaty rules that led to the inter-institutional disputes examined in chapter five, the vagueness of the Article 86 legislative procedure similarly provided wide scope for interpretation and resulted in an ECJ interpretation that was contrary to the Advocate General's opinion. The highly controversial nature of the decision prevented the ECJ from ruling formally until March 1991, almost 30 months after the motion had been filed, that the Commission's use of Article 86(3) as the Terminal Equipment Directive's legal basis was legal.[18] The ECJ could have conceivably ruled in favour of the Member States (in accordance with the 'objective' (Wincott 1995, p. 601) opinion of the Advocate General), but instead it ruled in favour of the Commission. This gives further support to the theory that the court maintains an institutional bias for integration.

Since the ECJ understood that the Member States did not oppose the substance of the directive, it was able to make an adverse decision against two powerful member states, France and Germany, with the knowledge that the governments would probably accept the decision. Highlighting the strategic behaviour of the ECJ, Garrett (1995, p. 178) points out that 'such cases are ideal from the court's perspective because they allow the ECJ to extend its authority – both in terms of the ambit of European law and its reputation for being able to enforce its decisions even against governments from powerful member states – without jeopardizing its legitimacy'.[19]

Meanwhile, all Member States were moving further down the liberalisation path. Indeed, the fact that a majority of Member States had largely implemented both the Terminal Equipment and Services Directives prior to March 1991 may also have influenced the ECJ ruling. Craig and DeBúrca (1995, p. 73), for instance, argue that the ECJ is susceptible to political pressure and that its 'judgments are clearly sometimes influenced by relatively 'non-legal' arguments … or by critical responses from national and from Community sources'. Moreover, the fact that Germany did not join the second legal challenge (against the Services Directive) implies that the German government had, in the meantime, tacitly accepted Commission authority to utilise Article 86. Likewise, Schmidt (1998, p. 175) notes that France may have opposed the Services Directive, not so much out of continuing opposition to Commission use of the legislative procedure, but more to be consistent with its opposition to the Terminal Equipment Directive (and thus to avoid losing face or weakening its position in that court case).

DG Competition officials also see the fact that the UK joined neither of the legal challenges as having been crucial to the success of the Article 86 directives. One DG Competition official commented that even an 'unbiased lawyer within the ECJ may have had problems agreeing with the Commission' without the UK's support.[20] Setting an important precedent, the ECJ ruled that the Commission could require the abolition of special or exclusive rights infringing the free movement of goods or services, the freedom of establishment, or the prohibition of abuses of a dominant position.[21] The abolition of exclusive rights concerning terminal equipment was justified on the grounds that these rights were incompatible with the principle of free movement of goods (Article 28 of the EC Treaty, formerly Article 30).[22]

Further clarifying the scope of inter-institutional authority, the ECJ went on to explain that the possibility that the Council might adopt legislation under the EC Treaty rules for positive integration (as examined in Part II) in one of the fields covered by Article 86, does not preclude the Commission from exercising its authority under Article 86. Both powers are complementary.[23] Although the ECJ nullified the provisions relating to special rights because the term had been insufficiently defined in the directive,[24] the overall balance of the ruling significantly enhanced Commission authority and advanced EC integration. The November 1992 ECJ ruling on the Services Directive further confirmed the Commission's authority.[25]

The Commission hailed these rulings as being both 'brave' (*European Report*, 20 March 1991) and 'an important constitutional judgment'.[26] Competition Commissioner Leon Brittan commented that 'doubts which have been expressed about the Commission's powers in this regard have now been removed and the Court has clarified an important area of law and policy' (*Agence Europe*, 21 March 1991). A Commission statement declared that the ECJ ruling had 'confirmed the Commission's jurisdiction to prohibit the maintenance of monopoly rights that are contrary to the competition provisions of the EC Treaty' (*EuroWatch*, 27 November 1992). As the *European Report* (18 November 1992) explained, these Court rulings are:

> highly significant insofar as it gives the Commission more power and allows it, if need be, to wield the Article in question again so as to break up other monopolies in the future. One example that comes to mind is the telephone sector: if the Commission should fail to gain satisfaction in this area through consultation with the Member States, it could trade in its velvet glove for an iron one in the form of Article 86 of the EC Treaty.

These two landmark ECJ decisions thus clarified institutional authority under the treaty competition rules. At the same time, the increased confidence of DG Competition officials would help facilitate their work in extending liberalisation.[27] Commissioner Brittan declared that 'armed now with judicial approval, it is inevitable that the Commission will increasingly consider the option of using Article (86)'.[28] Wincott (1996, p. 183) argues that the ECJ

> has been most effective when its rulings have altered the balance of power in the policy-making process so as to facilitate the passage of legislation which might otherwise have failed to become law. The interaction of the Court with other

institutions and interests in the development of Community law and in the policy process is a crucial characteristic of European integration.

Scharpf (1999, p. 58) contends that, as a result of successive ECJ rulings, 'the four economic freedoms, and the injunctions against distortions of competition, have in fact gained constitutional force *vis-à-vis* the member states'.

7.4   THE LIBERALISATION OF VOICE TELEPHONY SERVICES

Since Article 86 is subject to the Article 86(2) exception (that undertakings entrusted with a service of general economic interest are subject only to the Treaty rules in so far as these rules do not prevent the undertakings from performing their tasks), the Services Directive (preamble 18) had recognized that 'the opening-up of voice telephony to competition could threaten the financial stability of the telecommunications organizations ... [therefore] these restrictions are compatible with Article (86)(2) of the Treaty'. Despite considerable evidence to the contrary, national PTT Ministers had gained this exemption by arguing that it would be impossible to provide a universal service if competition were permitted in voice telephony services (which accounted for approximately 90 per cent of their revenue).

The Service Directive required the Commission to review this exemption in 1992.[29] In April 1992, Competition Commissioner Brittan noted that:

> Public voice telephony was excluded from the 1990 directive's liberalisation of the market because of its importance for revenue generation and the maintenance of universal service. The Commission should now reassess this position, given the growing need for more efficient intra-Community corporate communications and the vastly increased requirement for continent-wide networks in the new European environment. The liberalisation of public voice telephony services between the Member States and of supporting networks may contribute to meeting these needs and requirements.[30]

Following the ECJ's confirmation of the Commission's authority under EC competition rules, Competition Commissioner Brittan indicated that it was within the authority of the Commission 'in its own right'[31] to review the remaining exemption. 'It will itself implement existing competition rules'[32] and determine whether the restriction on the provision of voice telephony services remained justifiable.

In June 1992, however, Denmark rejected the Maastricht Treaty and the Community subsequently experienced a normative shift towards respecting the subsidiarity principle (see chapter six), which adversely affected the Commission's legitimacy and led to a loss of confidence. Shortly after the Danish vote, Commissioner Brittan proclaimed the importance of subsidiarity in EC activities and that it 'must be treated as a guiding political principle as well as a legal restraint'.[33] Henceforth, Commissioner Brittan consistently emphasised the need to consult extensively all interested parties, and especially the Member States, in reviewing the exemption. As discussed further below, it is clear that, after 1992, Commission officials toned down their approach in accordance with the changed 'logic of appropriateness' (March and Olsen 1989, p. 160).

In seeking to advance the liberalisation of the telecommunications sector, the Commission employed the services of an epistemic community to legitimate its desired policy line. As Haas (1992, p. 15) points out, 'epistemic communities can help formulate policies … In some cases, decision makers will seek advice to gain information which will justify or legitimate a policy that they wish to pursue for political ends.' The Commission thus commissioned three studies from external consultants that fed into the review process. National Economic Research Associates (NERA) produced a report, which concluded that there was significant potential for the realisation of additional long-term efficiency gains if competition was introduced into the voice telephony network. This was seen as the logical step in developing efficient and competitive European telecommunications networks for the benefit all users.[34]

A study produced for the Commission by Arthur D. Little Limited showed that by maintaining the *status quo* and simply following through with the initiatives taken previously, average annual telecoms growth rates would not exceed 4 per cent in Europe. In contrast, by pursuing a policy of increased and orderly liberalisation, the annual growth rate could rise to an average of 7 to 8 per cent.[35] A study produced for the Commission by Analysys Limited confirmed the overall conclusions of the Arthur D. Little study. It argued that any loss of market share for the incumbent TOs resulting from liberalisation would be more than offset by the growth in revenues and profits from the expansion of the telecommunications sector as a whole.[36] These studies enabled the Commission to conduct a more knowledgeable assessment of the exemption, clearly reinforced its preference for increased liberalisation, and armed the Commission with valuable information against the arguments put forth by the more conservative Member States.

Based on the three studies discussed above, the Commission drafted a report (1992c) on the situation in the telecommunications services sector (the 1992 Review report). In line with the increased sensitivity of EC institutions to Member State concerns generally, Commission officials postponed adopting the report until after the September 1992 French referendum on the Maastricht Treaty out of concern that the liberalisation proposals would put the French government in a difficult position (*Tech Europe*, 10 September 1992). The increased concern for subsidiarity also influenced the Commission's policy preferences. The Commission did not follow the policy recommendation put forth by its epistemic community of external consultants (see above). Their studies had concluded that the full liberalisation of voice telephony would achieve the most rapid Community growth rates in terms of economic expansion and the introduction of new services. Instead, in line with the views of the more conservative Member States, the Commission maintained that full liberalisation would have given rise to practical problems unless questions such as tariff rebalancing and access charges were resolved.

The Commission thus favoured the partial opening of voice telephony to competition. The 1992 Review report concluded that the liberalisation of voice telephony between Member States would not disrupt national systems, threaten the financial stability of the TOs, nor compromise universal service obligations. The Commission left open the question of whether future measures would be adopted by the Commission under Article 86 or by the Council under another legislative procedure. As an EC lawyer, Simon Taylor, commented:

> It may be that the legal proceedings arising from the Services Directive and Terminal Equipment Directive, although resolved largely in the Commission's favour, had a sobering effect on the Commission's ambitions for further Article [86] Directives and convinced the Commission to proceed on the basis of consensus with and between Member States.[37]

The normative shift towards respecting subsidiarity clearly left Commission officials less willing to assert their extensive institutional authority under the EC Treaty competition rules. The political capacity of the Commission to pursue its own agenda was thus weakened as the institutional balance of power shifted towards the Member States and the wider policy environment was more critical of both the Commission and European integration.

The insight of March and Olsen (1984, 1989), that organisational norms and values have a key role in determining the behaviour of institutional actors, helps account for the reluctance of Commission officials to embrace full liberalisation or Article 86. This pragmatic approach to liberalisation resulted from the heightened efforts of the Community institutions to respect the subsidiarity principle.[38] Commission officials contend that this normative shift occurred as a result of the 'Euro-scepticism' which built up in Denmark, France and other European countries 'where everybody started to say look, maybe it's dangerous to have all of this power moving to Brussels'.[39]

Maastricht thus 'marked the end of the "permissive consensus" on the part of the European publics with regard to European integration' and demonstrated the limits of the 'Monnet method' (Mazey 2001). The 'logic of appropriateness' (March and Olsen 1989, p. 160) had likewise changed, prompting Commission officials to pursue less aggressive liberalisation and seek a consensus for reform, instead of using the treaty's competition rules to force through their preferred policy line. In order to counter any potential objections based on subsidiarity, the 1992 Review report (Commission 1992c, p. 7) envisaged:

> minimum action necessary at Community level in order to remove obstacles to the provision of the widest possible range of telecommunications sources. Within the framework thus created at the Community level, Member States will continue to determine their own telecommunications policies.

In contrast to the policy-making process surrounding the Terminal Equipment and Services Directives, the Commission sought to initiate a pluralistic consultation process before drafting legislative proposals. The Review (p. 3) noted that:

> the Commission proposes to consult a wide range of interested parties, and in the first place the governments of the Member States, so that they have an opportunity to comment on the Commission's findings ... and all of these views can be taken into consideration before any final decisions are taken.

On a considerably more sympathetic note than in previous statements, Competition Commissioner Brittan was prepared 'to listen to the arguments of those resisting deregulation and even take up their demands if needed' (*Tech Europe*, 5 November 1992). Telecommunications Commissioner Pandolfi also emphasised the Commission's willingness to take account of Member States' concerns and potential deregulation difficulties. Thus, Luxembourg, where more than 25 per cent of telephone service income came from intra-EC calls (the EC average was about 4.4 per cent) was designated a 'special case which would have to be carefully examined by the Commission' (*Tech Europe*, 5 November 1992). Greece, Belgium, Ireland and Portugal, which received between 10 and 14 per cent of their telephone service income from intra-EC communications, 'stood to lose a lot of money from deregulation' and were also singled out for attention (*Tech Europe*, 5 November 1992). In early 1993, the new Commissioner for Telecommunications, Martin Bangemann, asserted that he would 'seek a consensus [on the liberalisation proposals] ... and did not favour using the Commission's full legal powers to pry open the market' (*CommunicationsWeek International*, 8 March 1993).

The definition of institutions under historical institutionalism is therefore found to be useful in examining the authority of political actors under the treaty rules on competition. Whereas 'rational choice' institutionalism focuses on the formal rules and procedures that enable and constrain political actors, the wider definition of institutions in historical institutionalism enables the researcher to consider the effects of understandings, routines, values and norms on the behaviour of political actors. Although the formal rules and procedures under the treaty rules for competition policy had not changed during the course of 1992, it is clear that the political context of the policy-making process had been altered.[40] The Commission was 'weakened with these [referendum] votes, with this Euro-scepticism'.[41] In accordance with a greater emphasis on subsidiarity and consensus, Commission officials submitted their liberalisation proposals to the Member States and affected interests for an extensive period of consultation.

Widespread Public Consultation and the Institutionalisation of an Intergovernmental Policy Network

Demonstrating once again the importance of British support for the Commission policy line, and partly out of concern for the changed attitude and reduced vocal support of Commission officials,[42] British Telecommunications Council President, Edward Leigh, sought to rally users to lobby directly both national governments and the Commission to further the liberalisation process.

> There is now a real opportunity to create genuinely and fully competitive telecoms markets both here in the UK, within Europe and globally ... I can assure you today that the Government intends to pursue [liberalisation] both here, within the EC and on the international scene ... But we should not underestimate the influence of some entrenched vested interests who in many cases are rehearsing arguments about the dire effects of telecoms competition which were proved to be wrong in the UK nearly a decade ago. It is important that this view is countered by users both directly with the Commission and also with the Governments in other Member States. The views of those vested interests must not be allowed to hold sway.[43]

In an attempt by Member States to claw back power from the Commission, the November 1992 Telecommunications Council created a powerful High Level Committee of National Regulators (High Level Committee) at the EC level. Established initially as an *ad hoc* committee to work with the Commission in carrying out the consultations on the Review, the High Level Committee gained a significant degree of influence over policy development. Analysys Consultants, a member of the Commission's epistemic community (see above), noted in a self-funded study (Analysys 1994) that the High Level Committee 'drove the Review process and defined the limits within which it would operate. After the [1992 Review report], the role of the Commission was to organise the hearings and report to the [High Level Committee] on the results.' The Member States thus gained a key role in formulating the liberalisation programme due to the reluctance of DG Competition officials to push through their desired policy line. An initial meeting between Commission officials and the High Level Committee in January 1993 defined the structure and focus of the public consultation process.

Demonstrating that the policy-making process for extending negative integration had become highly pluralistic, a series of public hearings over the liberalisation proposals were held in January and February 1993. These hearings attracted more than 130 organisations, including users and user associations; telecommunications operators; service providers; potential new entrants; equipment manufacturers; and trade unions. In addition, more than 80 written comments were received from individuals, companies and industry associations; users and user associations; ETNO and the majority of TOs individually; existing and potential service providers; manufacturing associations; equipment manufacturers; and the Joint Telecommunications Committee.[44] This trend towards pluralism is consistent with the Commission's general shift after 1992 towards more open and transparent policy making and the fostering of links with groups (Mazey and Richardson 2001a).

In a straightforward story of competing interests, INTUG argued for full liberalisation before January 1996.[45] The Union of Industrial and Employers' Confederations of Europe (UNICE) called for liberalisation of voice telephony 'as early as possible, aiming to achieve full liberalisation of all voice telephone services by the middle of this decade'.[46] BT, having already been exposed to competition in the British market for ten years (see chapter two), called for an Article 86 directive to require full liberalisation before January 1995.[47]

On the other hand, the TOs from the less liberalised Member States, including those from Greece, Italy, Portugal, Belgium, Luxembourg and Spain did not favour further liberalisation in the short term. Although accepting the ultimate inevitability of liberalisation in the long term, this group of TOs promoted the consideration of other options, including increased tariff regulation.[48] Meanwhile, France Télécom and Deutsche Telekom were concerned that the Commission proposal did not 'provide a clear long term perspective which allows TOs to develop their business and plan for investments along a predictable line of regulatory development'.[49] Although not opposed to further liberalisation, because competition was a recent phenomenon in their domestic markets (see chapter two), France

Télécom and Deutsche Telekom wanted to ensure that they had sufficient time to prepare for voice telephony liberalisation.

As with the policy formulation stage for positive integration examined in chapter four, EC level consultation mechanisms have increased the ability of interests to influence the formulation of liberalisation proposals. As Mazey and Richardson (1997, p. 195) point out, this demonstrates that groups have begun 'to play a significant role in the process of European integration, as predicted by the neo-functionalists. As such they now play a crucial role in the processes of problem identification and "framing", options search, and policy formulation within the Commission.' Illustrating once again the changed institutional setting resulting from changed norms, Commission officials noted that the consultation process was in line with the 1992 Birmingham declaration, which committed the Commission 'to consult more widely before proposing legislation which could include consultation with all the Member States and a more systematic use of consultation documents'.[50]

Following three further meetings between the Commission and the High Level Committee during February 1993, under which the Member States were able to give substantial input into policy formulation (see Commission 1993a), Commission officials put forth a two-stage draft liberalisation plan. Based on the preliminary results of the consultation, this plan proposed the liberalisation of intra-EC telephone services in January 1996 and the full liberalisation of the sector in January 1998. In order to accommodate those opposed to rapid liberalisation, Commission officials once again included the possibility of additional adjustment time for less-developed and peripheral regions. Thus, even under the treaty rules on competition, the Commission began to act as a broker, seeking to draft proposals that would be accepted by the Member States and affected interests (see Mazey and Richardson 1997, p. 179).

On 25 March 1993, Commission officials presented the draft plan to a Round Table for the Chairmen of the Network Operators in order to get their reaction. BT, well placed to take advantage of liberalisation, was very enthusiastic. Although somewhat less eager, the other TOs were prepared to endorse the Commission proposal. Only the Greek TO (OTE) and the Luxembourg TO (Administration P&T) clearly opposed the plan. OTE, as one of the least advanced (and least competitive) TOs, once again favoured strengthening tariff legislation. Administration P&T, on the other hand, wanted the plan to include a derogation for small networks (*Tech Europe*, 1 April 1993). Since Luxembourg was the only Member State that earned a majority of its income from intra-EC and international calls, Administration P&T contended that they faced losing considerable income or, perhaps, completely disappearing under the Commission's deregulation plan. In addition, nearly all the TOs opposed the deregulation of intra-EC services as a separate, intermediate stage. France Télécom, for example, did not feel it made sense to waste energy creating a complicated system that would last for only 24 months, but felt that the time could be better used to prepare for full liberalisation (*Tech Europe*, 1 April 1993).

A final meeting of the High Level Committee on 26 March 1993 enabled the Commission to take into account NRA concerns before forwarding the draft plan to the Council. Although some regulators were not enthusiastic about liberalisation, they all acknowledged its inevitability. Despite concerns expressed by the French, German and Belgian regulators that total deregulation in 1998 was overly optimistic (because preparations for liberalisation would take longer than planned), most regulators tentatively agreed on 1998 for complete liberalisation and opposed the intermediate stage. Since the British government is a rational self-interested actor, and the UK was the most advanced Member State in terms of liberalisation, the UK Minister in charge of telecommunications regulation was alone in favouring both the intermediate stage and complete liberalisation prior to 1998. The less advanced Member States, such as Greece, Italy, Portugal, Spain and Ireland, made their approval largely conditional on the receipt of Structural Fund assistance (*Tech Europe*, 1 April 1993). The less advanced Member States sought to maximise their gains from cooperation by negotiating a package deal. In demanding side payments for infrastructure development in exchange for a commitment to liberalise, they hoped to facilitate the achievement of a 'mutually acceptable and welfare-increasing solution' (Scharpf 1997, ch. 6).

Following the normative commitment of the EC institutions to respect subsidiarity, the Member States had been able to institutionalise a powerful intergovernmental policy network within the previously Commission-dominant area of negative integration. The High Level Committee enabled the NRAs to play a key role throughout the consultation process. In an instance where the Member States reversed the unanticipated consequences of Article 86, the High Level Committee:

> decided the contents of the communication to the Council and the European Parliament in April 1993. The Commission's power to propose legislation had thus been circumvented by the Member States. In effect, the Member States had dictated the outcome of the Review from the publication of the [Commission] communication in October 1992. (Analysys Consultants 1994, p. 9)

The extensive influence of the Member States in the formulation of the liberalisation proposal highlights a shift in institutional authority. Institutional change resulted largely from the changed environment and, in particular, increased domestic opposition to supranational EC authority. As the institutional balance of power shifted towards the Member States, DG Competition officials found themselves 'in a very defensive situation'.[51]

The changed environment altered the opportunities and constraints that DG Competition officials faced as their own authority under the treaty's competition rules had been curtailed. These shifts in authority can be frustrating at times to policy makers, 'as actors who have worked for years to gain an advantage over their competitors within a subsystem suddenly find their plans knocked awry by [external] events ... over which they have little control' (Sabatier 1987, p. 657). Likewise, DG Competition officials shifted their strategy. Indeed, exogenous changes can cause the strategies or goals that are being pursued within existing institutions to shift – or, in other words, outcomes can change as 'old actors adopt new goals within the old institutions' (Thelen and Steinmo 1992, pp. 16–17). Instead of pushing a directive through unilaterally, DG Competition officials sought to achieve a consensus of affected interests and national governments in favour of their liberalisation plan.

Following the extensive public consultation process, and after making final changes based on the comments of the TOs and NRAs, the Commission released a *Communication on the Consultation on the review of the Situation in the Telecommunications Services Sector* in April 1993, which highlighted the pluralistic nature of the review process that enabled affected interests to influence the proposal. The Commission (1993a, p. 3) made a virtue out of the consultation:

> the aim of the consultation was to obtain the view of both the European industry as a whole and of the main players in the telecommunications sector on the issues identified in the Commission's Communication of October 1992 ... Indeed, the oral and written comments received by the Commission helped it to refine and re-focus on the areas in which further action was most appropriate and urgent.

The Commission found that there was a general consensus that the liberalisation of telecommunications services was inevitable and necessary to meet technological and market demands. In line with the wishes of a majority of TOs and NRAs, as discussed above, the Commission concluded that the intermediate stage of liberalising intra-EC voice telephony was unnecessary. Instead, the Commission proposed the liberalisation of all voice telephony services (domestic, intra-EC, and international) from January 1998. Once again, illustrating the important role of the Commission as a broker seeking to balance the views of affected interests, the new Competition Commissioner Karel Van Miert explained that:

> the intermediate stage limited to the liberalisation of intra-Community traffic ... has not been favourably received by the majority of operators and regulators.

Moreover, certain of the users' representatives considered that this option had few advantages because the largest share of telephone communications are still made within the Member States. They believed that, for this reason, few new operators would be likely to enter a relatively small liberalised market. Equipment manufacturers consider the distinction between intra-Community and domestic traffic to be artificial. This is why I think that the Commission should not carry on with this proposition, despite the fact that it was seen in a favourable light last October. I feel that this demonstrates once again that the public consultation gave rise to a real exchange of views and that the Commission has taken account of the views expressed. Consultation is not a facade but a genuine dialogue that has influenced the decision-making process as it should.[52]

Commission officials also sought to put forth a balanced approach to infrastructure liberalisation. BT, Mercury and PTT Telecom Netherlands, having been exposed previously to competition in infrastructure, along with numerous service providers and users, argued strongly in favour of extending competition to infrastructure. Deutsche Telekom and France Télécom, with the support of a majority of TOs, on the other hand, voiced concern over the complicated nature of infrastructure liberalisation. They pointed out the implications infrastructure liberalisation could have for their ability to maintain financial equilibrium, continue long-term investment planning and supply a universal service. The Commission (1993a, p. 28) sought a compromise, in that it recommended the liberalisation of cable and alternative network infrastructure for already liberalised services before 1995, but as regards the full liberalisation of network infrastructure, 'the current consultations did not allow a final position to be formulated. Once additional experience is gained, a further broad consultation will be needed to determine the best way forward.'

The concept of 'policy style' (Richardson 1982) sheds light on the different ways that the Commission has utilised its competition authority. Richardson (1982, p. 13) defines policy style as 'the interaction between (a) the government's approach to policy-making and (b) the relationship between government and other actors in the policy process'. In trying to gain a handle on the Commission's policy style in the negative integration of the telecommunications sector, there was a noticeable shift away from the authoritative style of the late 1980s and early 1990s, under which the Commission relied on its formal authority under the competition rules of the EC Treaty to push through its desired policy line over the resistance of several Member States. Following the changed 'logic of appropriateness' (March and Olsen 1989), which stemmed from a heightened concern for subsidiarity in mid-1992, Commission officials proceeded on the basis of a more consensual or 'accommodative' (Mazey and Richardson 1997, p. 195) approach.

The April 1993 communication therefore left open the option of whether the Commission would adopt any further measures under Article 86. In presenting the proposal, Competition Commissioner Van Miert stressed the importance of

> preconsultation in the process which permitted a widespread consensus to emerge. It is not just an empty facade. This consensus means in the long run that the programme of liberalisation can be more effectively implemented rather than try an immediate big bang approach to liberalisation which only risks raising the fierce opposition of operators and delays the effective introduction of competition.[53]

Highlighting, once again, the increased role of the Member States in the formulation of the liberalisation proposal, the plan was forwarded to the Telecommunications Council in order to gain the formal approval of the Member States.

Following 'very long discussions'[54] in COREPER and two meetings of the Telecommunications Council, the Member States finally reached agreement on the liberalisation proposal in June 1993.[55] Council President and Danish Minister Arne Melchior perceived a common understanding among Member States that it was in their best interest to achieve liberalisation as rapidly as possible (*Agence Europe*, 18 June 1993). Although the Council confirmed the January 1998 date for voice telephony liberalisation, the extent that individual Member States were able to gain concessions to the liberalisation programme is demonstrated by the derogations granted. Luxembourg gained an explicit transitional provision for small networks, while Spain, Greece, Portugal and Ireland each obtained an automatic derogation of five years.

Eight delegations (all the Member States except France, Germany, the UK and the Netherlands) also expressed concern over the Commission proposal to liberalise alternative infrastructures before 1995 (*Agence Europe*, 18 June 1993). Infrastructure liberalisation was therefore not included in the Council resolution. Confirming once again the increased influence of the Member States on the liberalisation agenda, and highlighting the diminished informal agenda-setting capabilities of the Commission, Commission officials correspondingly removed infrastructure liberalisation, at least temporarily, from the Commission's agenda. According to one of the DG Competition (DG IV) officials involved in the negotiations, 'national governments rejected, or would not let through, anything on infrastructure ... [it] was really thrown out in force'.[56] The Member States nonetheless sought to institutionalise their key role in the formulation of negative integration policies. The Council invited the Commission to continue working with the High Level Committee on proposals 'aimed at strengthening the competitiveness of European operators'.[57]

7.5   CONCLUSION

The institutional setting provided for under the treaty rules on competition policy largely shaped the politics surrounding the negative integration of the telecommunications sector between 1988 and 1993. The Commission's extraordinary formal agenda-setting authority helped Commission officials achieve the bulk of their desired policy line.

DG Competition officials initially relied on their power in this institutional context, combined with an authoritative policy style, to push through the 1988 directive liberalising terminal equipment and the 1990 directive liberalising value-added services. Several Member States nevertheless contested the Commission's authority to issue these directives in the ECJ. Providing some support for the argument that the ECJ has an institutional bias for furthering integration, the ECJ upheld the Commission's right to issue directives unilaterally under the Treaty's competition rules. Following the normative shift to subsidiarity in 1992, however, Commission officials did not push the Commission policy line by resorting to, or even threatening to use, the treaty's competition rules to issue a directive unilaterally. A more consensual or 'accommodative' (Mazey and Richardson 1997) policy style enabled the Commission to gain the agreement of affected interests and the Member States on a time frame for the liberalisation of voice telephony prior to Commission formulation of a draft directive.

In line with historical institutionalism, therefore, this chapter finds that formal rules and procedures are not the only shaper of politics. Although the formal agenda-setting authority of the Commission to issue competition directives unilaterally had not been reduced, the policy-making process was altered and the informal agenda-setting authority of the Commission was diminished. The policy formulation process became more pluralistic and the balance of power shifted towards the Member States through the institutionalisation of the High Level Committee of National Regulators. The changed outcome resulting from this modified institutional environment is best demonstrated by the fact that DG Competition officials postponed their plans to liberalise telecommunications infrastructure in the face of strong concerns expressed by some national governments and affected interests.

NOTES

1. Competition Commissioner Karel Van Miert speech at the Royal Institute of International Affairs (Chatham House) on 'Analysis and guidelines on competition policy', London, 11 May 1993.
2. Case 155/73; (1974) ECR 409, (1974) 2 CMLR 177. See Pelkman and Young (1998), p. 145.

3. Before the 1988 Terminal Equipment Directive, the only other application of an Article 86 directive had been a 1980 directive on the transparency of financial relations between Member States and public undertakings (80/723, (1980) OJ L195/35). The Commission's authority to issue this directive was also challenged in the ECJ (Cases 188–90/80, *France, Italy, and the United Kingdom v. Commission* (1982) ECR 2545, (1982) 3 CMLR 144). The ECJ upheld the authority of the Commission. See Craig and De Búrca (1995), pp. 1068–1070 and pp. 1078–1079.
4. For a more detailed history of the EC's approach to competition policy, see Cini and McGowan (1998), pp. 15–37.
5. Interview, DG Competition official, Brussels, May 1999.
6. Interview, DG Competition official, Brussels, April 1999; interview, DG Competition official, Brussels, May 1999. Indeed, retrospectively, DG Competition officials still maintain that Article 86 is the only way that liberalisation could have been achieved as rapidly as it was.
7. Leon Brittan speech at the Centre for European Policy Studies 'Competition policy and Post and Telecommunications', Brussels, 19 May 1992.
8. Interview, DG Competition (DG IV) official, Brussels, July 1999.
9. Interview, DG Competition (DG IV) official, Brussels, July 1999; see also Thatcher (1997); Schmidt (1998); and *Handelsblatt*, 2 May 1988 as cited in Schneider and Werle (1990), p. 101.
10. Commission Directive of 16 May 1988 on competition in the markets in telecommunications terminal equipment (88/301/EEC) *Official Journal* L131, 27.05.88.
11. Interview, First Secretary Responsible for Telecoms, UK Permanent Representative, Brussels, May 1999.
12. These services, termed value-added or non-reserved services, included information and message forwarding services (for example electronic mail) and transaction services (for example financial transactions or telereservations).
13. Quoted in 'EC Commission waves red flag with provocative article 90 directive', LRP Publications, 30 June 1989.
14. CEC Press Release (1990) 'Dawn of a new era in European telecommunications', (IP 589), 18 July.
15. Council of Ministers, Conclusions of Meeting of 28 April 1989.
16. Commission of the European Communities, 'Commission Directive on competition in the markets for telecommunications services', 90/388/EEC, 28 June 1990.
17. Quoted in *Agence Europe*, 17 February 1990.
18. Case C-202/88: *French Republic v. Commission of the European Communities*, Judgment of the Court of 19 March 1991 (91/C 96/04).
19. Although Garrett (1995, p. 178) was referring to decisions 'against an *unimportant* sector in a powerful member state', (italics added) and the telecommunications sector is clearly an important sector, the fact that the German and French governments agreed (and had already largely complied) with the liberalisation intent of the proposals would similarly negate the political costs of implementing an adverse decision (or the potential political benefits of non-compliance).
20. Interview, DG Competition (DG IV) official, Brussels, May 1999.
21. Case C-202/88, Grounds 21–22.
22. Case C-202/88, Grounds 36 and 43.
23. Case C-202/88, Grounds 23–27.
24. Case C-202/88: *French Republic v. Commission of the European Communities*, Judgment of the Court of 19 March 1991 (91/C 96/04).
25. Joined Cases C-271, C-281 and C-289/90: *Kingdom of Spain and Others v. Commission of the European Communities*: Judgment of the Court of 17 November 1992, (92/C 326/08).
26. Quoted in the 'Business Guide to EC Initiatives', 1991 Spring, The EC Committee of the American Chamber of Commerce.
27. Interview, DG Competition (DG IV) official, Brussels, May 1999.

28. Leon Brittan speech at the Centre for European Policy Studies 'Competition policy and Post and Telecommunications', Brussels, 19 May 1992.
29. Preamble 22 and Article 10; Preamble 20 of the 1990 ONP Framework Directive also required the Commission to review the voice telephony and network infrastructure exemptions.
30. Leon Brittan speeches at the Business Week conference on the 'Future of world telecommunications', New York, 21 April 1992 and at the Centre for European Policy Studies 'Competition Policy and Post and Telecommunications', Brussels, 19 May 1992.
31. *Ibid.*
32. *Ibid.*
33. Leon Brittan speech at the European University Institute, (IP/92 1477, 92/06/11) 11 June 1992.
34. 'Study of the application of the ONP concept to voice telephony services', a report prepared under contract for the Commission by National Economic Research Associates, London (March, 1991).
35. 'Telecommunications issues and options 1992–2010': Study prepared for the Commission, Arthur D. Little, October 1991.
36. 'Performance of the telecommunications sector up to 2010 under different regulatory and market options', study prepared for the Commission, Analysys, February 1992.
37. Taylor, Simon (1993) 'Contributed article by Norton Rose [International law firm] on the outcome of the consultation stage of the European Commission's Telecommunications Services Review' in the *Reuter European Community Report*, 26 May 1993.
38. Interview, DG Competition (DG IV) official, Brussels, March 1999; interview, DG Information Society (DG XIII) official, Brussels, June 1999.
39. Interview, DG Competition (DG IV) official, Brussels, March 1999.
40. As soon as the Maastricht Treaty entered into force in November 1993, however, the subsidiarity principle was codified into EC law (see chapter six). March and Olsen (1989) would note that, to some extent, this formalized the 'logic of appropriateness'.
41. Interview, DG Competition (DG IV) official, Brussels, July 1999.
42. Interview, DTI official, London, May 1999.
43. Edward Leigh speech at Telecommunications Managers Association conference in Brighton 'Telecoms users urged to join battle against vested interests opposing international liberalisation', UK Government Press Release, 30 November 1992.
44. The corporatist joint Telecommunications Committee representing TO management and trade unions was set up by the Commission to assist in the formulation and implementation of Community policy. 'Commission Decision of 30 July 1990 setting up a Joint Committee on Telecommunications Services' (90/450/EEC).
45. INTUG (1993) 'Response to the Services Directive Review', January.
46. UNICE (1993) Position paper on 1992 Review, cited in 'Unice calls for further liberalisation', *Tech Europe*, 4 March 1993.
47. BT (1993) 'Choice for Europe's customers: British Telecommunications plc's response to the CEC 1992 Review', January.
48. ETNO (1993) 'ETNO report on the Review of the situation in the telecommunications sector (1992)', February.
49. *Ibid.*
50. Commission Press Release 'Commission launches Review of the telecommunications services sector', 21 October 1992 (IP/92/837).
51. Interview, DG Competition (DG IV) official, Brussels, May 1999.
52. Van Miert speech to the *Syndicat international des services postaux des telegraphes et des telephones* at the 'Telecommunications: The Way Forward' conference in Brussels, 15 April 1993.
53. Van Miert speech to the *Syndicat international des services postaux des telegraphes et des telephones* at the 'Telecommunications: The Way Forward' conference in Brussels, 15 April 1993.

54. Interview, DG Competition (DG IV) official, Brussels, May 1999.
55. Council Resolution of 22 July 1993 on the review of the situation in the telecommunications sector and the need for further development in that market (93/C 213/01).
56. Interview, Brussels, May 1999.
57. Council Resolution of 22 July 1993 on the review of the situation in the telecommunications sector and the need for further development in that market (93/C 213/01).

8. The full liberalisation of European telecommunications

INTRODUCTION

This chapter examines the negative integration of the European telecommunications sector that took place in the mid-1990s. It argues that the institutional setting provided for under the treaty's competition rules was an important factor in providing the Commission with the agenda-setting authority to achieve the full liberalisation of EC telecommunications.

This chapter is divided into two substantive sections. The first section demonstrates that an opened policy window facilitated Commission liberalisation efforts. In the face of mounting concerns over the development of the European Information Society, the Commission constructed a 'purpose-built' (Peterson and Bomberg 1999, p. 224) policy network to garner support for its desired policy line. At the same time, a shift in the preferences of two of the largest Member States (France and Germany) enhanced the Commission's informal agenda-setting capabilities. The second section analyses the Commission's efforts to take advantage of the more favourable political environment to coordinate the full liberalisation of the telecommunications sector. After briefly examining the Commission's efforts to liberalise both satellite and mobile communications, the final section investigates the Commission's efforts to achieve the full liberalisation of telecommunications infrastructures. This analysis shows that DG Competition officials were able to utilise, once again, a more authoritative policy style in order to pressure the reticent Member States into rapidly liberalising infrastructures.

8.1 THE INFORMATION SOCIETY: A POLICY WINDOW FOR COMMISSION LIBERALISATION EFFORTS

Having overcome Member State's concerns about subsidiarity, the Treaty on the European Union (TEU) was formally ratified in each Member

State and entered into force in November 1993. Title XII on trans-European networks, Article 129b of the TEU (now Article 154 of the EC Treaty) formally committed the EC to contribute to the development of telecommunications infrastructures; it aimed to enable EU citizens, market players and communities to benefit fully from the establishment of 'an area without internal frontiers'. Because explicit treaty reference to a policy area gives the Commission greater authority to propose legislation (Bulmer 1993, p. 365), the inclusion of telecommunications in the Maastricht Treaty enhanced the Commission's formal authority and was used to justify further action in the sector.

This increased authority, along with mounting concern that the emergence of the information society posed a threat to European competitiveness, gave new impetus to the liberalisation efforts: a window of opportunity appeared. In December 1993, the Commission turned its attention towards facilitating the development of the European information society in the *White Paper on Growth, Competitiveness and Employment: The challenges and ways forward into the 21st century*. One Commission official notes that, although the EC had been working on information technologies since the early 1980s (through the ESPRIT programme), this White Paper was the first time that the Commission sought to explain how these technologies fit into the EC's overall objective to establish a European information society.[1] Following the inability of the Member States to reach agreement on infrastructure liberalisation in June 1993 (see previous chapter), the White Paper was also the first Commission attempt to garner support for further telecommunications liberalisation. In seeking to jumpstart the liberalisation process, the Commission was hoping to regain its authority as an informal agenda setter, pushing forward the integration process by drafting 'new and innovative proposals which command the assent of the member governments and nudge the Union in a more integrative direction' (Pollack 1997, p. 121).

The White Paper (Commission 1993b) acknowledged that the emergence of an information society would be 'as important as the first industrial revolution'. Echoing the sentiments of Commission initiatives in the early 1980s, which had spurred the initial efforts at coordinating European telecommunications policies (see chapter three), the White Paper compared the EC approach to those initiatives underway in the US and Japan. The White Paper alerted the Member States to the major competitive advantages that would accrue to economies that develop the information society most rapidly. A condition (the importance of the information society) was thus defined as a problem through comparison with other countries (Kingdon 1995, p. 198).

Coupling a policy with the problem (Kingdon 1995, p. 201), the White Paper identified the liberalisation of the telecommunications sector as a necessary requirement for fostering the European information society. By linking the need for increased competition in telecommunications with the development of the information society, Commission officials also attracted attention from the Member States for their desired policy line. In reacting to the White Paper, the December 1993 European Council thus requested that the Commission organise a group of 'prominent persons' to draw up a report providing specific measures for action. Hence, the Commission had been able to set the agenda.

In response to the European Council's request, Commission officials constructed a 'purpose-built' policy network of influential actors at the 'very highest political level' to garner support for infrastructure liberalisation (Peterson and Bomberg 1999, p. 224). The High-Level Group on the Information Society, chaired by Telecommunications Commissioner Bangemann, was composed of 19 highly influential corporate officials, including former Commissioner Etienne Davignon (then President, Société Générale de Belgique), future Commission President Romano Prodi (then President, Institute for Industrial Reconstruction (IRI)), and 15 other Chief Executive Officers from major corporations including Olivetti, Siemens, Alcatel and Channel Plus. Following six meetings in early 1994, the group drafted a document, *Europe and the Global Information Society: Recommendations to the European Council* (the Bangemann Report), in May 1994 (High Level Group 1994).

In line with the interests of these private firms, the first line of the Bangemann Report urged 'the European Union to put its faith in market mechanisms as the motive power to carry us into the Information Age'. Arguing once again in favour of the same solution to the collective action dilemma, the first recommendation requested Member States to 'accelerate the on-going process of liberalisation of the telecoms sector by opening up to competition infrastructures and services still in the monopoly area'. As can be expected, the Commission supported the conclusions enthusiastically: it declared that the Bangemann Report put forth a convincing argument for communications and information to be at the centre of Europe's social and economic strategy.[2]

Confirming Mazey and Richardson's (2001) contention that private interests now play an influential role in formulating the Commission's policy line, a Commission official noted that the Bangemann Report was written by the group of corporate executives and merely endorsed by Commissioner Bangemann, who then had the Commission approve it.[3] This episode also highlights the way in which the Commission uses interest groups to endorse EC intervention, thereby legitimising its policy initiatives and ensuring support for those initiatives. Developments in the political sphere, such as the construction of this policy network, can also be a potent agenda setter (Kingdon 1995, p. 198). This powerful group of industrial actors helped the Commission regain its informal agenda-setting authority: it enhanced the credibility of the Commission policy line and facilitated DG Competition's efforts to extend liberalisation. A DG Competition (DG IV) official asserts that 'the impact on infrastructure liberalisation, what helped it, made it progress faster, was this Bangemann Report ... it's about the information society and so forth, but it helped to get it through'.[4]

In response to the Bangemann Report, the June 1994 European Council invited the Commission to establish a work programme of measures to facilitate the establishment of the European Information Society. The Commission accordingly developed a set of recommendations in *Europe's Way to the Information Society: An Action Plan*.[5] The Commission continued to gain ground as a policy entrepreneur and began to push its pet solution to the policy problem. As with previous initiatives, the Information Society Action Plan thus linked infrastructure liberalisation to the development of the information society; it pointed to 'infrastructure liberalisation as one of the main initiatives to be taken in order to open the way for the development of the network and applications on which the information society relies'.[6]

Two further developments in the political stream helped to push the liberalisation of infrastructure higher on to the EC's agenda. First, consistent with neofunctionalism, interest group support for further telecommunications liberalisation continued to swell. Outside of the Commission's formal consultation mechanisms (that is, the Bangemann Group discussed above), industry representatives increasingly pressured national governments for rapid action. In particular, the influential European Round Table of Industrialists (ERT), which 'enjoys access to European heads of government', (Grant 1993) initiated a campaign to lobby the Member States directly. The ERT includes representatives from about 50 of Europe's leading corporations and is widely regarded as being influential in pushing through the Commission's 1992 programme (see Green Cowles 1995).

The ERT formed a working group on Information Infrastructures in January 1993 that produced a report, *Building the Information Highways*. This report, which was sent to the heads of government before the June 1994 European Council, put forth a seven point strategy which 'underlined the fundamental interest that large European companies have in the opening of telecommunications markets, in view of the construction of the European Information Society'.[7] Later in 1994, the ERT sent a letter to each of the national telecommunications ministers (before the November 1994 Council meeting), and government leaders (before the December 1994 European Council meeting), urging them to achieve services and infrastructure liberalisation 'in the very near future. The proposed deadline of end 1997 is too far away … [because] European competitiveness is already at risk.'[8] In coordinating this campaign for increased liberalisation, and consistent with neofunctionalism, the ERT demonstrates the extent to which 'large firms [had] moved away from the 'national champion' strategies towards … a transnational solution to the future of wealth creation in Europe and to Europe's competitiveness *vis-à-vis* global trade' (Greenwood 1997, p. 248).

At the same time, sector-specific interests became increasingly organised at the European level. British, French, German and Scandinavian cable operators, for example, formed the Association of Private European Cable Operators (APEC) to promote competition in infrastructure in October 1994. APEC declared that 'unless cable networks are allowed to carry full telecommunications services, the citizens of Europe will miss out on a golden opportunity to benefit quickly and easily from the benefits of competition'.[9] The increased interest group focus on lobbying European leaders for further liberalisation served as an additional reform pressure. As Mazey and Richardson (1993, p. 249) have pointed out, 'the Europeanization of interests within (and beyond) the twelve Member States, can be … a potential force in the process of European integration'.

As both interest group and technological and economic pressures began to mount (see chapter 2), Member State preferences started to change; national governments began to accept, and even support, further liberalisation efforts (also see chapter 2). Thus, the policy window opened wider, as another key development in the political sphere further enhanced the Commission's ability to act as a policy entrepreneur. As Pollack (1997, p. 125) explains, for the Commission to be successful as an informal agenda setter,

> the distribution and intensity of preferences among the Member States is absolutely central. As in the case of formal agenda-setting, the Commission cannot simply propose its own ideal points without regard to the preferences of Member States, particularly when those preferences are clear and intense.

In this instance, the shift in Member State preferences was critical in enabling the Commission to set the agenda and coordinate further liberalisation.

Illustrating the dynamic relationship between EC-level and national developments, French and German governments (and their national TOs) increasingly supported EC-level developments during 1993 and 1994, as an additional justification for the pursuit of national reforms over domestic opposition (see chapter two). French and German governments were thus involved in a two-level game, seeking to increase their ability to manoeuvre domestically while helping to develop the EC's agenda. In other words, they were trying 'to maximize their own ability to satisfy domestic pressures, while minimizing the adverse consequences of foreign developments' (Putnam 1988, p. 434). In fact, by the end of 1994, both France and Germany had shifted from passively accepting the Commission policy line to actively supporting the rapid liberalisation of telecommunications infrastructures. In addition, the accession of three new Member States added two strong advocates of further liberalisation (Sweden and Finland) and only one that resisted liberalisation (Austria). Although these countries did not become official EU members until 1995, their representatives began participating in Council meetings towards the end of 1994. In any case, it is likely that Commission officials would have taken account of their preferences when formulating the liberalisation agenda.

A 'Window of Opportunity' is Opened for Commission Action

Having been removed from the Commission agenda (or at least moved down the list of priorities) in mid-1993 due to the strong opposition of most Member States and their national TOs (see previous chapter), infrastructure liberalisation attained a prominent position on the Commission agenda by the end of 1994. At the same time, the liberalisation of both satellite and mobile communications (which the Commission had been aiming to achieve for a few years) moved into better positioning for an authoritative decision to be made. As Kingdon (1995, p. 202) explains, 'the complete joining of all three streams – politics, problems, solutions – dramatically enhances the odds that a subject will become firmly fixed on a decision agenda'. The political developments discussed above (politics) had thus combined with mounting concern over Europe's place in the information society (problem) to provide the Commission with an optimal opportunity to pursue its pet solution: the full liberalisation of telecommunications services and infrastructure.

As the policy environment became more favourable, the Commission became more assertive in its approach to policy development. In particular, the Commission set the liberalisation agenda through the conscious pursuit of a twin-track – or carrot and stick – strategy. On the one hand, it endeavoured to mobilise a political consensus for reform through consultative Green Papers and negotiations with Member States and affected interests (the carrot). On the other hand, reflecting its stronger position, the Commission also increasingly adopted a more authoritative policy style to push through liberalisation proposals by invoking its formal powers under Article 86 of the EC Treaty (formerly Article 90) (the stick). The rest of this chapter demonstrates that the Commission's informal agenda-setting power was contingent on the more favourable political environment. Sections 8.2 and 8.3 briefly examine the Commission's exploitation of the policy window to achieve long-standing goals of reforming satellite and mobile communications. Section 8.4 examines, in greater detail, the Commission's more authoritative approach to infrastructure liberalisation; it demonstrates that the Commission's twin-track strategy was crucial for opening – and keeping open – the policy window.

8.2   THE LIBERALISATION OF SATELLITE COMMUNICATIONS

The deregulation of satellite communications proved to be difficult to achieve, due in large part to the existence and resistance of three powerful intergovernmental International Satellite Organisations (ISOs): EUTELSAT, INTELSAT and Inmarsat. Since the incumbent national TOs (the former PTTs) were the national signatories to the ISOs, they were able to control access collectively to approximately two-thirds of the available European space segment capacity required for satellite communications. The incumbent national TOs, therefore, were effectively able to coordinate supply between themselves and restrict access to potential competitors.[10]

Extensive consultation on a November 1990 Green Paper found widespread industry support for Community action to reduce the dominance of the ISOs, and thus the national TOs, over access to the space segment.[11] The UK, Ireland, Luxembourg, Italy, the Netherlands and Germany largely favoured the Green Paper reform recommendations. On the other hand, in line with the traditional French concern for maintaining sovereignty, as is manifest in its preference for the intergovernmental CEPT (see chapters three and six), France led the other Member States in opposing the liberalisation aspects of the Green Paper. As French PTT Minister Paul Quiles argued in 1990, there was 'no need for deregulation of the space sector, because the services offered by EUTELSAT are sufficient to cover user needs' (*European Report*, 14 November 1990). This resistance, along with the normative shift towards respecting subsidiarity in 1992, effectively prevented Commission officials from tabling a liberalisation proposal until the policy window discussed above opened for Commission action in 1994. In particular, the preferences of several Member States, including France (having joined the UK, Germany and the Netherlands in liberalising the satellite communications services market), had increasingly shifted towards supporting EC-wide liberalisation of the sector.

The Commission thus aimed to set the agenda through a June 1994 communication that argued for rapid restructuring of the sector.[12] Previous initiatives taken in the early 1990s had not successfully liberalised the space segment; a majority of the Member States still limited access exclusively to the national TO. The communication asserted that:

> despite the mandate to the Member States to undertake urgent actions as a result of the Satellite Green Paper of 1990 and the related Council Resolution of 1991, notably in broadening access to space segment of intergovernmental systems, only limited results have been achieved, and these only in some Member States.

Following public consultations that, once again, found widespread industry support for Commission action, the College of Commissioners planned to adopt an Article 86 directive in September 1994 under a written procedure.[13] Nonetheless, some of the less-liberalised Member States continued to oppose rapid liberalisation of the sector. In particular, Portugal demanded additional time to comply with the terms of the directive. Because the Portuguese government lacked the formal authority to veto the proposal, Portuguese Commissioner Joao de Deus Pinheiro delayed Commission adoption so that the Portuguese government could obtain a derogation for compliance (*Reuter European Community Report*, 27 September 1994).

In using his authority to protect Portugal's interests, Commissioner Pinheiro violated Article 213 of the EC Treaty (formerly Article 157 EC), which is supposed to ensure that the 'independence [of Commissioners] is beyond doubt'. At first glance, the Commissioner's violation of an EC Treaty rule appears to contradict new institutionalism, which posits that actors are constrained by institutions. A closer look at the two variants relied upon, however, demonstrates that they are sophisticated enough to explain the Commissioner's actions.

Rationalist choice theorists would point out that humans do not simply follow rules automatically. Instead, they are intelligent and have their own views and preferences, which can sometimes lead them to disregard the rules and norms they are supposed to adhere to (Scharpf 1997, p. 21). Thus, although the treaty's competition rules formally privilege some actors (European Commissioners) over other actors (national representatives), this approach also presupposes that European Commissioners are rational actors with subjective perceptions and preferences defined, for the most part, exogenously to the institutional setting. Rational choice institutionalists would, therefore, assert that Commissioner de Deus Pinheiro felt his interests were more in line with those of his native Portugal, and the Portuguese government, than in obeying his formal duties under Article 213. This approach implies that analysis should start with institutional explanations and then 'search for information on more idiosyncratic factors only when the more parsimonious explanation fails' (Scharpf 1997, p. 42).

Historical institutionalism, on the other hand, points out that institutions are principally normative in nature. For historical institutionalists, the normative consideration that Commissioners are not really expected to be totally independent from their national governments would, in this instance, more appropriately define the institutional setting than the formal EC Treaty rules. Indeed, Nugent (1994, p. 86–87) explains that 'full impartiality is neither achieved nor attempted ... Indeed, total neutrality is not even desirable since the work of the Commission is likely to be facilitated by Commissioners maintaining links with sources of influence.' Historical institutionalism, therefore, does not rely on institutions being formal structures, but rather a collection of norms, understandings, rules and routines. In disputing the rational choice assumption that political actors always act rationally, March and Olsen (1989, p. 22) contend that action is frequently based on the identification of behaviour that is normatively appropriate. To explain Commissioner de Deus Pinheiro's contradiction of the formal Treaty rules, March and Olsen (1989, p. 22) would point out that 'rules are codified to some extent, but the codification is often incomplete. Inconsistencies are common. As a result, compliance with any specific rule is not automatic.'

Thus, under both rational choice institutionalism and historical institutionalism, analysis begins with institutions, but become more sophisticated when idiosyncrasies or inconsistencies become apparent. Because of their sophistication, both approaches are capable of explaining Commissioner de Deus Pinheiro's apparent failure to act in accordance with the EC Treaty rules. In the end, the Commissioner achieved his goal. Following direct meetings between Portuguese Communications Minister Joaquim Ferreira do Amaral and Commissioner van Miert, they agreed upon a one-year derogation for the less-developed Member States. The Commission then redrafted and adopted the Satellites Directive unilaterally in October 1994.

Highlighting increased Commission confidence in its competition authority, Competition Commissioner Van Miert justified the Commission's use of the Treaty's competition rules: the 'use of directives based on Article [86] is an efficient tool for the application of competition law to such sectors, providing investment certainty to market agents and cutting red tape'.[14] In the post-Maastricht environment, however, Van Miert nevertheless acknowledged the need to use this authority scrupulously: 'this instrument must be used with care and in clearly circumscribed circumstances'.[15]

8.3   THE LIBERALISATION OF MOBILE COMMUNICATIONS

Prior to the opening of the policy window discussed in section 8.1, community action in mobile communications had also been stalled. In December 1991, *CommunicationsWeek International* noted that a Green Paper on mobile communications 'is already considerably delayed. The latest Commission date for publication is set for the second half of next year [1992].'[16] The 1992 Review contended that, although a general liberalisation trend for mobile communications had emerged (Denmark, Germany, Greece, France, Portugal and the UK had already liberalised, or were in the process of liberalising the sector), the possibility for extending the Services Directive to include mobile communications should be examined.[17]

Nonetheless, it was not until the policy window opened in 1994, that the liberalisation of mobile communications services gained a prominent position on the Commission's agenda. In April 1994, the Commission published a Green Paper on mobile communications based substantially on independent studies by external consultants.[18] In particular, a total of seven studies were produced specifically for the Green Paper, four of which were paid for by the Commission. This demonstrates the extent to which the epistemic community of telecommunications experts had become institutionalised within the Commission. As Haas (1992, p. 4) explains:

> the members of a prevailing [epistemic] community become strong actors at the national and transnational level as decision makers solicit their information and delegate responsibility to them ... To the extent that an epistemic community consolidates bureaucratic power within national administrations and international secretariats, it stands to institutionalize its influence and insinuate its views into broader international politics.

Attempting to place the liberalisation of mobile communications firmly on the EC's agenda, the first recommendation of the Green Paper was to amend the Services Directive to include mobile communications. Highlighting the fact that Member State influence had become institutionalised in the formulation of negative integration proposals, the Commission worked closely with the High Level Committee of National Regulatory Authorities and the Round Table for the Chairmen of the Network Operators in drafting the proposal. Demonstrating that the policy-making process for negative integration had also become highly pluralistic, public consultations attracted more than 250 organisations, including ETNO, service providers, equipment manufacturers, INTUG, ECTUA and other users.

Although most contributors agreed that the liberalisation of the sector should proceed immediately, institutional inertia prompted most former

PTTs to act as they had in the past (Krasner 1984, p. 235) and, therefore, they opposed competition in mobile infrastructure. Once again, despite substantial evidence to the contrary, they maintained that their ability to provide a universal service would be hindered. Individuals within the PTTs may have also resisted change for fear that it would result in a loss of status, policy scope and budgetary support (Krasner 1984, p. 235). Thus, the Commission (1994a) found that 'there was broad support for early action on liberalisation and direct interconnection, with the exception of a majority of existing fixed network operators'.

This resistance is significant in that it demonstrates that the TOs still maintained some effective veto power over far-reaching Commission initiatives: 'It's very clear this is not politically acceptable', explained Leo Koolen, a Commission official in charge of mobile communications. 'They [the TOs] believe they need a number of years to prepare their companies for full liberalisation of infrastructure.'[19] In addition, some NRAs wanted to avoid pre-empting the debate over a forthcoming Green Paper on infrastructure liberalisation (see below).

Taking advantage of the more favourable political environment, the Commission recommended that mobile services be liberalised before 1996 through an amendment to the Services Directive. Nevertheless, as a compromise to the less-liberalised TOs, and demonstrating that the Commission did not seek to abandon entirely its accommodative policy style, the Commission ruled out immediate mobile infrastructure liberalisation. This was to be assessed within the 'global approach' (Commission 1994a) developed in the Green Paper on infrastructure liberalisation (as discussed below). Thus, notwithstanding the willingness of Commission officials to rely on the Treaty's competition rules to liberalise satellite communications and mobile services, they continued to act as policy brokers, limiting the amount of political conflict to acceptable levels and coming to a reasonable solution for the problem (Sabatier 1987, p. 662).

8.4   THE LIBERALISATION OF TELECOMMUNICATIONS INFRASTRUCTURES

As the previous chapter explained, during the 1992 Review consultations, the Commission created a consensus among Member States in favour of a timetable to liberalise voice telephony services. At the same time, however, little progress was made on infrastructure liberalisation. The opened policy window discussed above, combined with its formal agenda-setting authority under the Treaty's competition rules, enabled the Commission to adopt a more authoritative policy style after 1994 to push through proposals to liberalise telecommunications infrastructures expeditiously. Thus, although Commission officials broker policies, they also advocate policies, especially when they have a mission that is well defined (Sabatier 1987, p. 663). DG Competition's institutional mission to ensure the application of the EC Treaty rules on competition (see section 7.1 of previous chapter) compelled DG Competition officials to take advantage of the favourable political environment to pursue infrastructure liberalisation more aggressively.

In contrast with the consensual policy style pursued during the 1992 Review consultations (when Commission officials refrained from advocating use of the Treaty's competition rules), towards the end of 1994 Commission officials began to declare their willingness to 'to force through [liberalisation] proposals if ministers fail to agree them over the coming months' (*Agence France Presse*, 27 October 1994). Competition Commissioner van Miert explained that Article 86 may be necessary to ensure that liberalisation takes place as rapidly as possible. Meanwhile, Telecommunications Commissioner Bangemann continued to tie together the problem with its solution (Kingdon 1995, p. 202). He warned that jobs could be lost if Europe failed to keep pace with the US and Japan (*European Information Service*, 7 October 1994): 'the information revolution has speeded up considerably. Economic as well as technical reasons require a rapid and efficient liberalisation.'[20]

Even though Commission officials were prepared, if necessary, to rely on their formal agenda-setting authority to push liberalisation proposals through unilaterally, they still preferred to set the agenda informally and attempt to broker a general consensus in favour of liberalisation. This is not surprising because it enabled them to avoid conflict, facilitate implementation and avoid – in the post-Maastricht political context of Euroscepticism – criticism of the Commission's power. Commission officials thus put forth a consultative Green Paper on the liberalisation of infrastructure (Infrastructure Green Paper) in October 1994.[21] Initially planned for December 1994, the Commission expedited drafting the Green Paper so that the November 1994 Telecommunications Council could examine it (*Agence Europe*, 13 October 1994). Aiming to obtain a consensus of Member States in favour of its proposal, the Commission hoped the Green Paper would enable Ministers to establish a clear liberalisation schedule.[22]

The proposals put forth in the Infrastructure Green Paper demonstrate that, during the months following the June 1993 Council agreement to liberalise services fully (see previous chapter), policy-oriented learning had taken place among DG Competition officials. Such learning within a system of beliefs poses little problem as advocacy coalition members are constantly attempting to understand better causal relationships and variable states that are in accordance with their policy approach (Sabatier 1987, p. 678). DG Competition officials had become increasingly aware

198 The liberalisation of European telecommunications

that it would not be possible to have a significant degree of competition in telecommunications services without simultaneously liberalising the corresponding infrastructure to provide those services. American officials had pointed out to DG Competition officials that the telecommunications sector was susceptible to the process of spillover. They had warned that if the national TOs maintained monopolies over infrastructure, potential service providers would face strong resistance to access to the public network. One DG Competition official explained that:

> I can remember a meeting some months [after the Member States agreed to the full liberalisation of telecommunications services] with some Americans, from the FCC [Federal Communications Commission] or the Department of Justice, and they told us that [liberalising services only is] very weak because you can't just provide a service if you have no access to infrastructure. In the end, you will be endlessly stuck in all these discussions on the price and access to infrastructure. They were right ... we needed to liberalise the provision of infrastructure to have real competition.[23]

Based on this learning, Commission officials sought to establish a reframing of the policy issue in the Infrastructure Green Paper. Further liberalisation efforts, therefore, were to be based on the principle that, as telecommunications services become exposed to competition, so should the infrastructure for their delivery. The Infrastructure Green Paper thus proposed a two-stage liberalisation schedule. The first stage would immediately introduce competition into infrastructure for the provision of already-liberalised services. A second stage, in January 1998, would result in full infrastructure liberalisation.

Despite the fact that the distribution of Member State preferences had become more favourable for Commission action (see section 8.1 above), Commission officials were nonetheless unable to get the Member States to agree on the two-stage liberalisation proposal at a November 1994 Telecommunications Council meeting. Germany, France, the UK and the Netherlands (supported by the accession candidates Sweden and Finland) favoured the Commission's plan. These six countries issued a declaration that urged the Commission to proceed as quickly as possible with proposals for the early liberalisation of alternative infrastructure. The British Minister, Ian Taylor, pointed out that 'we are not encouraging the Commission to use Article [86], but are absolutely determined to see rapid liberalisation of the alternative infrastructures'.[24]

The other eight Member States (supported by Austria), on the other hand, remained firmly opposed to the first stage of the liberalisation schedule (*European Report*, 19 November 1994; *Agence Europe*, 19 November 1994). As a result, the German Minister, Wolfgang Boetsch, chairing the meeting, settled for a compromise agreement on a single stage. The Council Resolution simply acknowledged the general principle that infrastructure liberalisation should take place before 1 January 1998. Seeking to maintain a key role for Member States in the formulation of liberalisation proposals, the Council invited the Commission to work with the High Level Committee of National Regulators. As with previous liberalisation proposals, the Council concluded that Spain, Portugal, Greece and Ireland should have an additional five years to comply and 'very small networks' an additional two years.[25]

Cable Infrastructure Liberalisation

The strong support of three largest Member States (the UK, France and Germany), along with the Netherlands, Sweden and Finland, in favour of the two-stage liberalisation proposal bolstered the Commission's position. DG Competition officials accordingly increasingly invoked their formal agenda-setting authority in order to push through infrastructure liberalisation over the resistance of the less-liberalised Member States. The following month (December 1994), the Commission issued a draft Article 86 directive aimed at liberalising cable television networks for the provision of already liberalised telecommunications services (Cable Directive) by January 1996. As Table 8.1 demonstrates, there was widespread scope for Commission action.

Table 8.1   *Use of cable television networks for telecommunications*
          services (as of January 1995)

Belgium	No
Denmark	No
France	Non-voice services only
Germany	No
Greece	No cable TV networks
Ireland	No legal provision
Italy	No cable TV networks
Luxembourg	No legal provision
Netherlands	Limited use
Portugal	No
Spain	No
UK	Yes
Austria	No
Finland	Yes
Sweden	Yes

Source: 'L'impact de l'authorisation de la fourniture de services de télécommunications liberalisés par les câblo-opérateurs', IDATE, 1994 as cited in Commission (1995) 'Cable Infrastructure Green Paper, part II'.

National government support for, or opposition to, the proposed directive demonstrates the self-interested nature of the Member States. Of these that supported the proposal, only Germany had not previously liberalised the use of cable networks for the delivery of value-added (non-voice) telecommunications services. Nonetheless, Deutsche Telekom was the only cable operator in Germany and therefore would not face increased competition as a result of the draft proposal. In those Member States that did not have a legal provision within their national regulatory framework, the legal ambiguity prevented cable TV networks from carrying telecommunications services.[26] The Commission proposal thus clearly sought to move the liberalisation process forward in a majority of Member States.

The proposal continued to link a solution (cable infrastructure liberalisation) to a problem: American market players had a significant advantage over their European counterparts because equivalent high-capacity infrastructure in the United States was ten times less expensive. The opened policy window prompted the Commission to declare its pet proposal as 'without doubt, the most important gateway to the "information society"'.[27] In line with Mazey and Richardson's (2001, 2001a) contention that private interests have begun to play a key role in problem identification and policy formulation, Commissioner van Miert noted that the draft directive 'respond(ed) to the specific requests'[28] of small and medium companies. Even though the Commission planned to adopt the proposal under its own authority, DG Competition officials recognised the increasingly pluralistic nature of the policy-making process for negative integration and the importance of the pluralistic network for effective implementation. DG Competition officials thus made the proposal available for widespread consultation with Member States, the EP and affected interests.[29]

Infrastructure Green Paper Consultation Process

Simultaneously, and in conjunction with the draft Cable Directive consultation, an extensive consultation took place over the Infrastructure Green Paper.[30] Interested parties were encouraged to submit written comments and attend a series of hearings during the first half of 1995 to enable the Commission to develop a package of proposals for widespread reform of the regulatory framework.[31] This initiative attracted responses from more than 200 organisations, businesses and associations.[32] Contributors included the traditional policy network surrounding EC telecommunications legislation: the TOs, NRAs, service providers and user and consumer organisations. The NRAs maintained their institutionalised role in the High Level Committee of National Regulatory Authorities, while the TOs also maintained a degree of institutionalised influence through the Round Table for the Chairmen of Europe's Network Operators.

Nevertheless, in this policy sector – as in others – as the range of EU policy has expanded, a wider range of interests have realized that EU policy is an important aspect of their institutional environment (Mazey and Richardson 1996, pp. 206–207). Accordingly, Commission proposals for negative integration have attracted an increasingly wide range of interests: the Infrastructure Green Paper consultation received contributions from the motion picture, broadcasting, data protection, insurance and tourism industries. Thus, the extension of EC telecommunications policy further confirms the logic of negotiation that exists between affected interests and the Commission.

Predictably, users, equipment manufacturers, industry, mobile operators and some TOs with experience in liberalised markets argued that the Commission should take immediate steps to liberalise alternative and cable infrastructures for the provision of already liberalised services. They saw this as essential to improving the efficiency of European industry (including TOs) and keeping Europe at the 'forefont of the Information Society'. ERT estimated that their collective $2.5 billion annual telecoms budget would be half that in the US; mobile operators emphasised their need to control costs better through access to alternative infrastructures; the *Bureau Européen des Unions de Consommateurs* (BEUC) contended that domestic consumers would benefit from more services and lower prices; while the Union of Industrial and Employers' Confederations of Europe (UNICE) and some operators argued that the local loop[33] needed to be liberalised for significant competition to occur: the proposed Cable Directive was seen as a key move in the right direction. APEC asserted that the removal of interconnection and voice service restrictions would ensure necessary cable networks investment, thus facilitating the development of the Information Society.

In contrast, most TOs and some NRAs wanted the Commission to delay infrastructure competition until 1998. They argued that the delivery of a universal service and, particularly, funding for the development of the public network could be destabilised. This was because their most profitable areas of business (which had been funding the universal service) were also the most likely to incur competition. Commission officials, however, disputed the accuracy of this view. As policy brokers, they sought to eliminate unsound data and causal assertions (Sabatier 1987, p. 680), by pointing out that the experience of liberalised markets, along with Commission studies, implied that the chance that limited infrastructure competition would threaten universal service provision was smaller than some TOs and NRAs feared. Commission officials also suggested that adequate safeguards could be established that were proportional to the proposed liberalisation.

On the other hand, consistent with its institutional bias for furthering integration, the Commission 'attach(ed) considerable weight to the calls from industry, business, service providers, users, and from mobile network operators with experience of competitive markets, for faster but limited infrastructure liberalization now'.[34] The Commission therefore aimed to set the agenda informally; it proposed the liberalisation of cable networks and alternative networks for mobile communications services before January 1996; and the full liberalisation of infrastructure from January 1998. Before, however, relying on its formal agenda-setting authority under the EC competition rules to push through the directives unilaterally, the Commission still hoped to gain a consensus of Member States in favour of its proposals.

The Lack of Member State Consensus for Rapid Infrastructure Liberalisation

Nonetheless, although the June 1995 Telecommunications Council agreed to liberalise telecommunications infrastructure fully prior to January 1998, the Member States continued to disagree over the need to set an earlier deadline for partial liberalisation. This disagreement demonstrates that, if the Commission had sought to adopt the proposals under the Treaty rules examined in Part II of this book (or a similar Treaty rule which would have required a qualified majority of Member States in favour of the proposal), EC-wide infrastructure liberalisation would have been delayed. Numerous interviews undertaken for this book, including those with Member State representatives, Commission officials, MEPs and TO and user group representatives, confirm the view that the liberalisation of EC-wide infrastructure progressed more rapidly because of the Commission's ability to resort to Article 86.

For example, and as discussed above, the early liberalisation of cable networks was supported by only six of the fifteen Member States: the UK, France, Germany, the Netherlands, Sweden and Finland (see also *Tech Europe*, 5 May 1995). 'Positions in fact are very divergent both in form (Article 86) and content ... of the draft directive' (*Agence Europe*, 15 June 1995). Because the Member States were unable to find common ground at the June 1995 Telecommunications Council, the conclusions were limited to asking the 'Member States which so wish to forward their detailed comments as soon as possible to the Commission' while calling 'upon the Commission to give the greatest consideration to these positions'.[35]

If the draft Cable Directive had been processed under the normal legislative procedures (see Part II), it most likely would not have been adopted. Indeed, with only the above mentioned Member States voting in favour, the proposal would have fallen 20 votes short of the 62 votes necessary (under the QMV rules) for Council approval. Nevertheless, the Commission proceeded to utilise its institutional authority under the Treaty competition rules to adopt the Cable Directive unilaterally in October 1995. Under the terms of the directive, EC cable networks were to be liberalised immediately for the provision of telecommunications services, although derogation was granted for the less-advanced and smaller Member States.[36]

Member State governments also continued to disagree over whether or not to allow, at an earlier date, the provision of already liberalised telecommunications services over alternative networks (that is, utility and railway networks). As discussed above, the UK, France, Germany, the Netherlands, Sweden and Finland favoured this approach, while the other Member States, led by Austria, Belgium, Portugal and Greece, continued to oppose it (see also *Reuter European Community Report*, 12 June 1995). As a result of the lack of consensus, the June 1995 Telecommunications Council Resolution failed to mention anything about alternative infrastructures.[37] The following month (July 1995), however, Commission officials once again chose to rely on their formal agenda-setting authority, under the Treaty competition rules, to issue a draft Article 86 directive liberalising alternative infrastructures from January 1996 (the Full Competition Directive).[38]

Similarly, the UK, France, the Netherlands, Sweden, Finland and Ireland were the only Member States to support the Commission proposal to liberalise mobile infrastructures prior to January 1996 (*Agence Europe*, 13 June 1995). The June 1995 Council Resolution therefore merely provided for 'the deployment by Member States of the best efforts to put in place at the earliest opportunity a scheme for the free establishment and use of infrastructure by mobile and personal communications services operators'.[39] Once again, therefore, the proposal failed to attract sufficient support required for Council adoption under the normal legislative procedures: these six Member States totalled only 35 of the 62 votes necessary under QMV. One week after the June 1995 Telecommunications Council meeting, the Commission therefore chose again to rely on its formal agenda-setting authority, under the treaty competition rules, to issue another draft directive. This Article 86 proposal aimed to liberalise the mobile telecommunications market (the Mobile Directive) before January 1996 (*Multinational Service*, 11 July 1995).

Increased Commission Reliance on its Competition Authority

In choosing to adopt these proposals under Article 86 of the EC Treaty, therefore, the ability of Commission officials to achieve their desired policy line was enhanced. As Weaver and Rockman (1993) point out, differences in policy-making procedures can have an influential effect on policy outcomes. In a comparison between several systems of government, they find (p. 12) that 'differences in decisionmaking processes give [some] governments ... greater capabilities to perform a variety of policymaking tasks ... [which] give them a better prospect of turning their policy choices into policy outcomes consistent with those choices'. The current case study clearly demonstrates that, by proceeding under the Treaty's competition rules, Commission officials increased the chances that they would be able to bring about the rapid liberalisation of cable, mobile and alternative infrastructures.

DG Competition officials also relied on their competition authority to get the French and German governments to reach agreement on a deadline for liberalising their alternative infrastructures before an EC-wide liberalisation timetable had been established. In particular, the institutional setting established under the 1990 Merger Control Regulation (MCR) enhanced Commission authority to regulate mergers. The MCR complemented its existing competition powers and provided competences that the Commission had sought unsuccessfully for more than a decade (Armstrong and Bulmer 1998, p. 114).[40]

During the middle of 1995, DG Competition officials used this increased authority to scrutinise a proposed joint venture between Deutsche Telekom and France Télécom. Because each of the national TOs controlled more than 70 per cent of their national data transmission markets, DG Competition officials initially disapproved of the alliance (*European Information Service*, 28 September 1995). However, after extensive negotiations, the French and German governments agreed to liberalise alternative infrastructures fully by July 1996. France Télécom and Deutsche Telekom also agreed to delay integration of their data transmission services until January 1998. DG Competition officials thus conditionally approved the joint venture.

In justifying Commission authority to examine such mergers, Competition Commissioner van Miert explained that:

> we are confronted with more and more arrangements of co-operation, joint ventures and mergers. While such strategic alliances are necessary in order to allow enterprises to reach the critical size for competitiveness on the world market, they must not foster monopolistic situations. Thus, liberalisation is a prerequisite for alliance, and the correct application of the competition rules remains one of the major tasks of the European Commission.[41]

A degree of autonomous supranational authority is demonstrated by the Commission's ability to pressure France and Germany into agreeing on a date to liberalise alternative infrastructures prior to the setting of an EC-wide deadline. Likewise, the Commission demand to postpone the integration of data services was not in line with the perceived best interests of their national TOs. Because data services integration would have accounted for about 80 per cent of the joint venture's revenue, the two companies accepted this delay 'with regret'.[42]

In other words, DG Competition officials were not simply responding to Member State preferences. In ensuring that the liberalisation of the French and German telecommunications markets was not impeded by an alliance between their national TOs, DG Competition officials were acting more in line with their normative commitment to promote EC-wide competition (see section 7.1 of the previous chapter). In this instance, DG Competition's preference did not reflect as much the intentions and preferences of the Member States – the principals – but instead the preference, and autonomous agency, of the Commission itself (Pollack 1997a, p. 107). This is crucial, because an important part of historical institutionalism is being able to demonstrate that institutions have an autonomous role in policy making. Indeed, the approach requires that 'political institutions are more than simple mirrors of social forces' (March and Olsen 1989, p. 18).

Having gained the agreement of France and Germany to liberalise infrastructures partially in 1996, and having already set the agenda formally by issuing draft Mobile and Full Competition Directives, it is clear that DG Competition officials invoked an increasingly authoritative policy style in order to ensure that EC-wide infrastructure liberalisation would take place expeditiously. At the same time, Commission authority was enhanced by a shift in the preferences of the fourth largest Member State: Italy began to support the liberalisation proposals actively. In July 1995, Competition Commissioner van Miert asserted that several Member States, Italy being the most recent, had requested the Commission to issue the directives unilaterally (*Agence Europe*, 20 July 1995).

Similar to France and Germany (see chapter two), the Italian government was also involved in a two-level strategy and sought to use EC-level developments to gain impetus for domestic reform.[43] Towards the end of 1995, Italian Telecommunications Minister Gambino was particularly supportive of unilateral Commission action.[44] Gambino had produced an ambitious draft reform law that remained blocked within the Italian Parliament; he therefore hoped that EC-level developments would facilitate his efforts at domestic reform. In accordance with Moravcsik's liberal intergovernmentalism, the EC can thus have a significant influence on domestic politics. The EC mantle adds weight within domestic debates over large reforms and strengthens the ability of national politicians to set the domestic agenda (Moravcsik 1994, p. 71).

Meanwhile, Denmark began planning for full liberalisation in mid-1995 and was keen to pursue further reform.[45] By late 1995, the preferences of the

Danish government accordingly shifted towards supporting the Commission liberalisation package. In fact, Denmark, supported by the UK, Finland and Sweden, urged the Commission to accelerate the liberalisation process. Denmark submitted a motion to the November 1995 Telecommunications Council which requested the Commission to bring forward the deadline for full liberalisation by two years. The Danish declaration sought to speed up the liberalisation of telecommunications infrastructure for already liberalised services, which included bringing forward from January 1998 to January 1996 the liberalisation of voice telephony services (*Financial Times*, 30 November 1995).[46] As Table 8.2 demonstrates, by the end of 1995, a majority of Member States favoured Commission action.

Table 8.2   Summary of Member State preferences for the early liberalisation of alternative networks and mobile communications (as of November 1995)

Austria	Reluctant
Belgium	Reluctant
Denmark	**Supportive**
Finland	**Supportive**
France	**Supportive**
Germany	**Supportive**
Greece	Reluctant
Ireland	Supportive if given extension to 2000
Italy	**Supportive**
Luxembourg	Some reluctance due to small size
Netherlands	**Supportive**
Portugal	Reluctant
Spain	Reluctant
Sweden	**Supportive**
UK	**Supportive**

Nonetheless, Member States such as Spain, Greece, Austria, Portugal and Belgium were less liberalised and feared that increased competition would significantly reduce the market share of their national TOs. They therefore continued to oppose the draft proposals and called on the Commission to slow down the liberalisation process. Telecommunications Council President and Spanish Minister Jose Borrell Fontelles was among the strongest critics, protesting that 'the way the Commission is jumping in, we could end up with sort of a banana republic, where the Commission simply brings out directives and that's the end of that'.[47] Similarly, Belgian Telecommunications Minister Elio di Rupo contended that 'in the workings of the Commission there is a situation which is beginning to worry us. Each Minister has to take political clout vis-a-vis the electorate. The Commission does not respect the delays called for by the Council, but takes decisions based on Article [86].'[48] A Portuguese official complained that the Commission was 'legislat(ing) themselves. What is the system in place in Europe? I don't understand this anymore.'[49]

Demonstrating, once again, the importance of the institutional environment in enabling the Commission to achieve its desired policy line, the proposals would probably not have been adopted under normal legislative procedures. Under QMV rules, Spain, Portugal, Greece, Austria and Belgium would have had enough votes to form a blocking minority. In proceeding under the treaty competition rules, however, the Commission ensured that these five Member States could not veto the draft directives. In other words, DG Competition officials utilised the Article 86 legislative procedure because it has fewer 'veto points'[50] (Immergut 1992) and therefore, is less vulnerable than Article 95 of the EC Treaty (formerly Article 100a). In enabling the Commission to adopt the directive without the formal approval of the Member States or the European Parliament, the Treaty's competition rules structured the power relations, helped shape the strategies of political actors and determine policy outcomes; Commission officials pursued infrastructure liberalisation more forcefully and achieved their goals more easily.

DG Competition's more aggressive approach is highlighted by the fact that the draft directives included harmonisation provisions despite the expressed opposition of nearly every Member State. For example, the draft Full Competition Directive established basic principles for licensing new entrants, introduced measures necessary to safeguard the universal service and specified that interconnection rules must be set down on a non-discriminatory basis before 1997. Representatives from almost every Member State, including the UK, France, Germany and Denmark, asserted that these provisions should have been dealt with in separate ONP legislation under Article 95 of the EC Treaty (formerly Article 100a). 'We find it okay to liberalise using Article [86],' commented a Danish official, 'but the detailed harmonisation elements we find problematic. We have sent comments about this to the Commission.'[51]

Nevertheless, in including these controversial harmonisation aspects within the proposals, Commission officials demonstrated the extent to which the logic of appropriateness had changed. In other words, the Commission was once again willing to push through its pet policy solution over Member State opposition. The balance of power had therefore shifted back towards the Commission, as DG Competition officials were somewhat less concerned than they had been during 1992 and 1993 (see previous chapter) with obtaining a consensus in favour of their liberalisation proposals.

In seeking to justify their proposals against Member State opposition, Commission officials contended that the disputed provisions merely established basic rules to ensure the progress of liberalisation. According to one Commission official, 'I don't think it makes sense to say liberalise, but not set ground rules to make it possible'.[52] 'I say that if we liberalise in 1998' remarked Competition Commissioner Van Miert, 'we need interconnection measures, we need the universal service to make it stick. Otherwise we take a decision without any practical consequences.'[53] At the November 1995 Telecommunications Council meeting, 'in the face of strong pressure from seven EU Member States to slow down its liberalisation programme ... Van Miert stuck to his guns' (*European Information Service*, 5 December 1995). The Commission's more aggressive approach is highlighted by Commissioner Van Miert's assertion to the Council: 'I left no ambiguity that everything would go ahead as scheduled'.[54] In early 1996, the Commission proceeded to adopt both the Full Liberalisation and Mobile Directives.[55]

8.5   CONCLUSION

This case study demonstrates the contingent nature of the Commission's informal agenda-setting power. With regard to the liberalisation of the telecommunications sector, the Commission, acting as a policy entrepreneur, consciously pursued a twin-track – or carrot and stick – strategy. Thus, as the policy environment became more favourable, the Commission became more assertive and invoked its formal powers under Article 86.

However, throughout the process, the Commission also endeavoured to mobilise a political consensus behind its proposals among both the affected interests and member governments, which was also important in opening – and keeping open – the policy window. This strategy is important not just in the context of this particular case study. More generally, it demonstrates awareness on the part of the Commission of the need to take account of the wider political context – the telecommunications infrastructure liberalisation which took place just after Maastricht, subsidiarity, and the public backlash against the EU in general, and the Commission in particular. Hence, the Commission at this time was generally keen to demonstrate its willingness to consult and to accommodate Member State difficulties.

With the appearance of a policy window in 1994, DG Competition (DG IV) officials were once again able to rely on a more authoritative policy style to achieve progress in areas where previous liberalisation efforts had either stalled or failed. The Commission's formal agenda-setting authority under the treaty's competition rules enabled DG Competition officials to formulate and adopt directives unilaterally to liberalise satellite communications, mobile communications and telecommunications infrastructures. Apart from the derogations granted to the less-developed countries, the Article 86 policy-making process foreclosed the opportunity for the Member States to dilute the liberalisation intent of the proposals.

The Commission's extraordinary authority to enforce the treaty's competition rules at least partly accounts for the Commission's success at coordinating the negative integration of European telecommunications during the 1990s. Indeed, most analysts, officials, and market players consider EC telecommunications liberalisation to have been a success. Many highlight the crucial role that the Commission has played in the liberalisation process.[56] Cini and McGowan (1998, p. 169), for example, conclude that 'despite some difficulties, Commission efforts to liberalise the European telecoms industry have largely been successful'. The Commission's fifth implementation report in November 1999 also found that the 'transposition of the liberalisation directives has been completed by all Member States with the exception of Portugal and Greece, which are due to liberalise fully on 1 January 2000 and 31 December 2000 respectively' (Commission 1991a, footnote 1).

The power relationships under the Treaty competition rules contrast with those relationships uncovered in the policy-making process surrounding positive integration (see Part II). At least partly because the Council has formal authority to 'water down' legislative texts in the inter-institutional bargaining stage, the Member States were able to implement the harmonisation directives in diverse ways, particularly in relation to national interconnection and licensing regimes (see chapter six). The bulk of the Commission's fifth implementation report therefore exposed the barriers that remain to a harmonised EC telecommunications market. Similarly, Iain Osborne (2000) concluded that liberalisation 'has been a decided success. Harmonisation, on the other hand, has been less successful.' Indeed, most market players, Commission and national officials interviewed for this book agreed that there was a significant lack of harmonisation in important aspects of the national regulatory frameworks following implementation of the ONP regulatory framework.

In line with the new institutionalist assumption that institutions can have an independent influence on policy outcomes, the institutional setting established under the EC Treaty competition rules was at least partly responsible for the Commission's comparatively greater success during the 1990s at coordinating liberalisation, as opposed to the harmonisation of EC telecommunications, under the ONP regulatory framework (1998 package).

NOTES

1. Interview, DG Information Society (DG XIII) official, Brussels, March 1999.
2. EC Press Release (P/94/60) 'Infrastructure: an essential step on the path to the Information Society', 26 October 1994.
3. Interview, DG Information Society (DG XIII) official, Brussels, March 1999.
4. Interview, Brussels, May 1999.
5. 'Europe's way to the Information Society: an action plan', Communication from the Commission to the Council and the European Parliament and to the Economic and Social Committee and the Committee of Regions (COM(94) 347 final) Brussels 17 July 1994.
6. EC Press Release (P/94/60) 'Infrastructure: an essential step on the path to the Information Society', 26 October 1994.
7. Author's own translation from the *Reuter European Community Report*, 16 November 1994.
8. ERT Letter to the Prime Ministers, Telecommunications Ministers, and Commissioners on infrastructure liberalisation, quoted in the *Reuter European Community Report*, 16 November 1994.
9. Richard Woollam, APEC representative and Britain's Cable Television Association Director General, quoted in *Reuter European Community Report*, 12 October 1994.
10. For a more detailed examination of the Community's satellite communications policy, and the dominant position of the ISOs, see the Commission's 'Communication on satellite communications: the provision of – and access to – space segment capacity', COM (94) 210 final, 10 June 1994.
11. Green Paper on a common approach in the field of satellite communications in the European Community (COM(90)490, 20.11.90).
12. Commission (1994) 'Communication on satellite communications: the provision of – and access to – space segment capacity', COM(94) 210 final, 10 June 1994.
13. For an explanation of the Commission's written procedure for adopting proposals, see Edwards and Spence (1997), p. 40.
14. Commission Press Release, 14 October 1994 (IP/94/948).
15. *Ibid.*
16. *Communications Week International* 'Europe struggles to define itself: EC telecoms liberalization seen as too slow, too limited', 16 December 1991.
17. 1992 Review of the situation in the telecommuncations sector, European Commission Communications SEC (92) 1048 Final.
18. 'Towards the personal communications environment: Green Paper on a common approach in the field of mobile and personal communications in the European Union', COM(94)145 final, 27 April 1994.
19. Koolen, Leo at the Brussels Telecommunications Forum, quoted in the *Reuter European Community Report*, October 10, 1994.
20. Commissioner Bangemann, quoted in EC Press Release 'Liberalising telecommunications infrastructure: an essential step on the path to the information society', (P/94/60) 25 October 1994.
21. Commission Green Paper, 'A common approach to the provision of infrastructure for telecommunications in the European Union', COM(94)682, 25 October 1994.
22. EC Press Release (P/94/60) 'Liberalising telecommunications infrastructure: an essential step on the path to the Information Society', 26 October 1994.
23. Interview, DG Competition (DG IV) official, Brussels, July 1999.
24. Quoted in *Agence Europe*, 19 November 1994.
25. Council Resolution of 22 December 1994 on the principles and timetable for the liberalization of telecommunications infrastructures, (94/C 379/03).
26. Green Paper on the Liberalisation of Telecommunications Infrastructure and Cable Television Networks, Part II, A Common Approach to the Provision of Infrastructure for Telecommunications in the European Union, European Commission, COM(94) 682, January 1995.

27. Commission Press Release 'The Commission opens cable TV networks to liberalised telecoms services – a first step to the multi-media world' (IP 94–1262), 21 December 1994 and *European Report*, 24 December 1994.
28. *Ibid.*
29. *Ibid.*
30. In addition to the two-stage liberalisation schedule put forth in the October 1994 Green Paper, an additional part of the Green Paper was put forth in January 1995 considering issues such as universal service and interconnection arrangements.
31. European Commission Press Release 'Liberalising telecommunications infrastructures: publication of part II of the Green Paper and consultation on the future regulatory framework' (IP/95/61), 25 January 1995; also see *European Report*, 28 January 1995.
32. Communication on the Consultation on the Green Paper on the Liberalisation of Telecommunications Infrastructure and Cable Television Networks, COM(95)158 final, 3 May 1995.
33. The local loop is the last half mile of telephone line, which is connected to individual consumers and businesses.
34. 'Communication on the Consultation on the Green Paper on the Liberalisation of Telecommunications Infrastructure and Cable Television Networks,' COM(95)158 final, 3 May 1995.
35. Conclusions of Council Meeting, Council of Ministers, COM Press Release, PRES 95-175, 13 June 1995.
36. Commission Directive 95/51/EC of 18 October 1995 amending Directive 90/388/EEC with regard to the abolition of the restrictions on the use of cable television networks for the provision of already liberalised telecommunications services, OJ L 256/49, 26 October 1995.
37. Council (1995) Resolution of 18 September 1995 on the implementation of the future regulatory framework for telecommunications, 95/C 258/01; OJ C 258/01, 3.10.95.
38. In addition, the directive sought to codify the previous Member State agreements to liberalise all telecommunications services (see previous chapter), and all telecommunications infrastructures (see above), prior to January 1998. See Commission Press Release (IP: 95-765) 19 July 1995.
39. Council Resolution on the further development of mobile and personal communications in the European Union, 95/C 188/02, 29 June 1995.
40. See, in particular, chapter four on 'The regulation of mergers and acquisitions in the single market' for a historical institutionalist account of the development of Commission authority in mergers.
41. European Commission Press Release (P/94/60) 'Liberalising telecommunications infrastructure: an essential step on the path to the Information Society', 26 October 1994.
42. 'EU nears approval of Atlas,' *UPI*, 16 October 1995.
43. For a more detailed account of the interaction between Italian and EC-level reform efforts, see Thatcher (1999b, 2004).
44. Interview, Italian Permanent Representation official, Brussels, June 1999.
45. See 'Denmark: gearing up for competition', in *Telecommunications (International Edition)*, October 1995.
46. The text of the Danish declaration is quoted in Multinational Service, 5 December 1995.
47. Quoted in the *Reuter European Community Report*, 24 November 1995.
48. Quoted in *Multinational Service*, 5 December 1995.
49. Quoted in the *Reuter European Community Report*, 24 November 1995.
50. A 'veto point' is a point in a policy-making process where opposition can mobilise and thwart policy development.
51. Quoted in the *Reuter European Community Report*, 24 November 1995
52. Quoted in the *Reuter European Community Report*, 24 November 1995.
53. Quoted in 'Despite pressure Van Miert sticks to his guns on liberalisation,' *European Information Service*, 5 December 1995.

The liberalisation of European telecommunications

54. *Ibid.*
55. Commission Directive 96/2/EC of 16 January 1996 amending Directive 90/388/EEC with regard to mobile and personal communications; and Commission Directive 96/19/EC of 13 March 1996 amending Directive 90/388/EEC with regard to the implementation of full competition in telecommunications markets, OJ L 74/13, 22 March 1996.
56. For the most far-reaching claim of the Commission's authority to push through the liberalisation directives, see Sandholtz (1998). For more balanced views, see Schmidt (1998), which acknowledges the Commission's need to take into account the views of the Member States and Thatcher (1999b), which highlights Commission pressure as only one of several pressures pushing Member States towards further liberalisation.

PART IV

Towards a new regulatory framework and theoretical approach

9. Towards the 2003 electronic communications framework

By the turn of the millennium, it had become clear that Europe needed more focus and a sense of urgency to catch up in the information society.[1] Information Society Commissioner Erkki Liikanen, September 2002.

INTRODUCTION

The chapter examines the development of the 2003 electronic communications framework. The first part of the chapter relies on Kingdon's (1995) approach to analyse the high level of uncertainty that is present during the initial stage of policy making. Following unsuccessful Commission attempts to establish a European Regulatory Framework (ERA) and to coordinate national licensing conditions under the ONP (Open Network Provision) regulatory framework, as discussed in chapter 6, and in response to the convergence of the telecommunications, information technology and audiovisual sectors, a new cycle of EC policy making began. One of the Commission's objectives was to increase EC-wide coherence, in areas such as the licensing of telecommunications operators, where previous harmonisation efforts had failed. Kingdon's approach, which focuses on streams of problems, politics and ideas, highlights the Commission's ability to take advantage of a policy window to pursue its pet policy solution: a modified institutional and regulatory framework more fully incorporated into the EC's supranational framework.

The second part of this chapter relies on a synthetic approach to analyse the development of the 2003 electronic communications framework. As was the case with the development of the 1998 package, no theory on its own could adequately account for the construction of the new framework. A synthetic approach, on the other hand, that applies multiple policy-making theories to the super-systemic, systemic and sub-systemic levels of analysis, helps to account for the multi-level character of the EU policy-making process, which is 'heavily nuanced, constantly changing and even kaleidoscopic' (Peterson and Bomberg 1999, p. 9). As has been demonstrated throughout the book, technological developments have been a consistent exogenous pressure for change in European telecommunications and, to some extent, have driven policy development. In anticipation of continuing technological advancement, the 2003 electronic communications framework is lighter, more flexible, technology-neutral and applies to all transmissions networks in the same way, so that it can encompass new, dynamic and unpredictable markets.

The final part of this chapter finds that, due to the deficiency of regulatory authority at the European level, the 2003 electronic communications framework establishes several committees of NRAs to coordinate implementation of the framework. In line with the thesis put forward by Eberlein and Grande (2005), these transnational regulatory networks are necessary because of the increased harmonisation of telecommunications policies at the European level and the corresponding failure to establish a European regulator to oversee national implementation. In particular, the 2003 electronic communications framework establishes a European Regulators Group to facilitate coordination between the NRAs because, as discussed in chapter six, the Member States successfully resisted Commission efforts to create a European Regulatory Authority.

9.1   A NEW POLICY-MAKING CYCLE: TOWARDS A NEW REGULATORY FRAMEWORK

Following implementation of the ONP legislation discussed in Part II of this book, a new cycle of EU policy development began. Institutional learning, or the 'learning process of how implementation by the national authorities corresponds to intentions' (Armstrong and Bulmer 1998, p. 56) took place and resulted in a Commission attempt to correct implementation gaps, such as the patchwork of distinctive national licensing regimes that remained despite the harmonisation intent of the 1997 Licensing Directive.

In the initial policy-making stage, numerous ideas are floating around for further policy development and for the establishment of the future regulatory environment. These ideas, including the reorganisation of the European-level institutional environment, can be explained with the garbage can model of decision-making (Cohen, March and Olsen 1972). In this model, policy makers, problems, solutions and choice opportunities come together as a result of being simultaneously available. The flow of each of these streams is autonomous.

> Although not completely independent of each other, each of the streams can be viewed as independent and exogenous to the system. Problems, solutions, decision makers, and choice opportunities are linked in a manner determined

by their arrival and departure times and any structural constraints on the access of problems, solutions, and decision makers to choice opportunities. (March and Olsen 1989, p. 13)

John Kingdon's (1995) version of the garbage can model, as presented in chapter one, is the most useful analytic framework for examining this early stage of policy making and the streams of problems, politics and policies.

Problem Stream

Kingdon (1995, p. 198) explains that 'officials learn about conditions through feedback' and that this feedback can be formal, as in 'routine monitoring' or 'evaluation studies', or informal, as in 'streams of complaints'. Here, Commission officials learned about problematic conditions through feedback, such as the Commission's implementation reports, ETO studies and complaints from market players, which uncovered deficiencies in the regulatory framework. Indeed, the seventh implementation report in November 2001 found 'continuing divergences in implementation' and that the overall situation was 'very disappointing'.[2]

Faced with this dissatisfaction in the performance of the ONP regulatory framework, the Commission was led to 're-examine its strategy' (Sabatier 1987, p. 675). As a result, the Commission concluded that there was a need to adjust the institutional and regulatory environment. The Commission (2000, p. 3) felt that the difference in licensing regimes across the Member States 'is not in line with the policy objective of stimulating the development of a competitive and dynamic market in communications services'. Pointing to the importance of encouraging the development of pan-European services, the Commission (2000, p. 3) contended, 'the existing divergence does nothing to help the process'.

In addition to a lack of licensing harmonisation, and as discussed in chapter 6, the Commission (2000, p. 4) concluded that the CEPT had been ineffective at coordinating EC telecommunications policies: 'The mechanism for harmonisation of licensing conditions and procedures through the CEPT … as foreseen in the current licensing Directive has not been successful. Even the minimal harmonisation format, the one-stop-shopping procedure, has failed to materialise.' There was increased uncertainty among political actors about the role of the CEPT in the future regulatory environment. For example, user group representatives did not feel that ECTRA continued to have a role following the completion of the ONP regulatory framework.[3] An OFTEL representative (and ECTRA member) argued that ECTRA is not very good at:

making quick decisions that are actually of any use. I think if you could shut down [ETO], it's a waste of time, and just concentrated on ECTRA being a body that attempts to spread best practice across the whole of the European continent ... we'd regard that as a worthwhile job and one that it actually does quite well at the moment. But [ECTRA] is large, and unwieldy, and cumbersome ... it's one of these lovely dinosaurs that's trying to find a role for itself, but hasn't quite got there yet.[4]

Commission officials also concluded that there were no underlying objectives in ECTRA policy. 'Both the CEPT and ... ECTRA are struggling with an identity problem caused by the changed telecommunications regulatory environment in Europe and the rapid market and technology changes in the information society sector which they are ill-equipped to follow'.[5] Since there is no automatic mechanism for eliminating an institution such as ECTRA, when it is no longer needed, history is often inefficient (Pierson 1996).

Commission officials thus sought to frame the problem, as they perceived it, in order to increase their ability to set the reform agenda. As Kingdon (1995, p. 198) explains, 'problem recognition is critical to agenda-setting ... policy entrepreneurs invest considerable resources bringing their conception of problems to officials' attention, and trying to convince them to see problems their way. The recognition and definition of problems affect outcome significantly.' Accordingly, Commission officials sought to publicise their view that the CEPT is no longer an adequate institutional forum for coordinating EC-wide re-regulation (see Commission 2000; DG Information Society 2000; Commission 1999).

Politics Stream

Events in the politics stream flow according to their own rules and their own dynamics (Kingdon 1995, p. 198). According to rational choice institutionalism, if the institutional framework produces policies that are in equilibrium, all actors can only worsen their existing benefits by institutional change. However, if the institutional framework does not produce policies in equilibrium, further reform and instability can be expected. Due to widespread dissatisfaction among Commission officials and other institutional actors, the European level institutional framework fell into '*dis*equilibrium' (Pollack 1996, p. 437). Instability and self-interested bargaining over reform ensued.[6] In this situation, an institution 'survives primarily because it provides more benefits to the relevant actors than alternate institutional forms' (Hall and Taylor 1996, p. 945). In the political stream, Commission officials and CEPT officials promoted the utility of their respective organisations.

CEPT officials argued that the usefulness of the CEPT lies in the fact that it covers substantially more European countries than the EU. Therefore, they asserted that the added value of the CEPT is the exchange of views between the eastern European countries and the Member States. Because of this, CEPT management felt that they should continue to organise pan-European coordination in cooperation with the EC.[7] Commission officials, on the other hand, countered that the EU already consults with the EEA countries and, with the expansion of the EU, more eastern European countries participate.[8]

At the same time, historical institutionalism highlights the normative shift of the EC towards increased concern with financial accountability. This trend was, at least in part, the consequence of increased EP scrutiny over the Commission's spending on studies during the late 1990s, which was reinforced by the collective resignation of the Commissioners in 1999. Subsequent to this, Commission officials expanded their efforts in scrutinising ETO work orders. A Commission official explained that, 'until recently the Commission has been very forthcoming towards ETO. The more critical approach of the last few months, partly induced by our internal financial control department ... has come as a surprise to ETO and ECTRA'.[9] Commission officials argued that the Commission would have to treat ETO work orders with more rigour and in line with the principles of sound and efficient financial management. They also acknowledged the need to establish a maximum of procedural transparency and ensure a more consistent and active management of the studies. Learning, or, the 'relatively enduring alterations of thought or behavioral intentions that result from experience' (Sabatier 1987, p. 672), took place as Commission officials recognised the need to be more cautious with financial expenditures. As a result of this learning, and the diffusion of these new beliefs and attitudes among Commission officials (Sabatier 1987, p. 672), the Commission's official position was transformed.

As Kingdon (1995, p. 198) explains, 'Developments in this political sphere are powerful agenda setters'. In particular, the new environment of financial accountability brought about by EP scrutiny over Community spending forced the Commission to take more concern in the results of its studies and caused the Commission to be more critical of its relationship with the CEPT.

Policy Stream

In the 'policy primeval soup', there is a tremendous amount of uncertainty (Kingdon 1995). Telecommunications Director-General Robert Verrue has pointed out 'the need to adapt certain rules stems from the changing environment. The telecom market is in particular affected by a rapidly changing parameter as regards the ongoing development of digital technologies, increasingly global competition, and the emergence of convergence.'[10] Continuing the strategy developed in the late 1970s (see chapter three), and consistent with the 'path-dependent evolutionary model', which 'implies that the effects of epistemic involvement are not easily reversed' (Adler and Haas 1992, p. 373), Commission officials employed the services of an epistemic community to study aspects of the sector and produce policy proposals.

Independent external consultants produced nine reports for the Commission in preparation for a comprehensive review of the regulatory framework in 1999. One of these studies examined the role of an ERA. The consultants that produced this study found that the 'political climate' did not allow for an ERA and, as a result, that there was little support for it (Watkinson and Guettler 1999). The changed preferences and aims of the group of actors that had previously favoured an ERA was largely due to increased fear of a long period of regulatory uncertainty, resulting from the institutional and political barriers discussed in chapter 6, if the Commission proposed an ERA formally. The study found that 'most operators/service providers are working for short-term solutions to immediate problems. They envisage a long period of political debate and haggling prior to the establishment of any EU regulator with indefinite results to follow some time in the future ... Major issues relating to the principle of subsidiarity will be raised.' The Commission (1999, p. 51), therefore concluded, 'improvement of existing institutional arrangements at the EU level will be more effective than setting up a completely new European regulatory institution'.

Commission officials maintain that the ERA study was an important factor in determining the Commission's policy line.[11] Commission support for the adaptation of the current institutional setting (instead of forming an ERA) thus highlights the influential role that the epistemic community has under high degrees of uncertainty. It also demonstrates that 'some ideas are selected out for survival while others are discarded, order is developed from chaos, pattern from randomness' (Kingdon 1995, p. 200).

Under conditions of uncertainty, political actors 'are prepared to engage in a negotiative process even where there is considerable disagreement' (Richardson 1996a, p. 13). CEPT and Commission officials negotiated over the future arrangements between their two organisations. Commission officials identified the problems discussed above and put forth ideas for reforming the CEPT/Commission relationship. Commission officials proposed institutional modifications to the CEPT aimed at preventing ECTRA modification of ETO work orders; giving the EC greater authority through the increased use of Commission mandates to ECTRA; and increasing implementation effectiveness through the mandatory reporting and monitoring of ECTRA decisions to the Commission.

The CEPT came to the Commission for ideas, put forth its own ideas and tried to restructure. In 2001, it eliminated ETO and merged its responsibilities into the European Radiocommunications Office (ERO). It also eliminated ECTRA and the European Radiocommunications Committee (ERC), and merged their responsibilities into a new organisation, the Electronic Communications Committee (ECC). The CEPT also tried to reorient its activities toward policy development and to encourage strategic thinking. For example, the second day of the July 1999 Kiev plenary was devoted to the 'strategic topic' of 'voice over the Internet'. Nonetheless, in contrast to the dominant role it played from the late 1950s until the late 1970s in coordinating European telecommunications policies, as examined in chapter three, the CEPT strategy had become one of damage limitation and the search for a future role.

In the policy stream, many ideas are put forth and floated around as numerous people try to influence the future regulatory framework. As was demonstrated in chapter four, the early stage of the EC policy-making process is highly pluralistic. Since 1997, the Commission has been publishing its internal working documents, draft legislation and green papers on its website along with the public comments received. Following its 1999 Review Communication, the Commission received more than 200 comments from NRAs, user groups, trade associations, industry and individuals, while a two-day public hearing in January 2000 attracted more than 550 participants. Sabatier (1987, p. 680) suggests that these forums are intended:

> to force debate among professionals from different belief systems in which their points of view must be aired before peers. Under such conditions, a desire for professional credibility and the norms of scientific debate will lead to a serious analysis of methodological assumptions, to the gradual elimination of the more improbably causal assertions and invalid data, and thus probably to a greater convergence of views over time concerning the nature of the problem and the consequences of various policy alternatives.

Although uncertainty is a key feature of this early stage of policy formulation, institutional analysis is valuable in highlighting the organisational disposition of garbage cans. March and Olsen (1989, p. 29) point out that garbage cans are affected by the structural features of the institutions with which they occur and can produce a bias. The Commission acts with a bias in its preference for increased EC authority. Commission officials did not leave the consideration of their pet proposals (for example, reform of both the Commission/CEPT relationship and national licensing regimes) 'to accident. Instead, they push(ed) for consideration in many ways and in many forms' (Kingdon 1995, p. 201).

The Commission sought to establish a common framing of its reform ideas through widespread discussions on institutional issues. It proposed new committee procedures within the supranational EC to assume the responsibilities of the CEPT and contended, 'new institutional arrangements are necessary' (Commission 1999, p. 51). In addition, it argued that national licensing regimes require 'adjustment at the level of the EU regulatory framework ... an efficient and effectively functioning single European market can be achieved by rigorously simplifying existing national regimes using the lightest regimes as a model' (DG Information Society 2000).

The NRAs, as can be expected, continued to support the CEPT role in EC telecommunications policies (Commission 2000a). The CEPT argued that:

> the strengthening of the co-operation between the EC and CEPT seems ... highly suitable for the benefit of the whole communications sector ... [and] recommend(s) to the EC [to] build upon the Committees and working groups of CEPT rather than trying to establish parallel structures within the EU, which may duplicate activities. (CEPT 2000)

EC members of the CEPT nevertheless acknowledge that 'where action within CEPT is regarded as unsatisfactory, EU action ... may yet prove more effective' (CEPT 2000).

Choice Opportunity and Policy Windows: the Joining of the Streams

Following the 1998 deadline for implementation of the ONP regulatory framework, the Commission was required to review national regulatory frameworks and propose further measures. The Licensing Directive (Commission 1997), in particular, obliged the Commission to prepare, before January 2000, a report to the EP and Council on the 'need for further development of the regulatory structures as regards authorizations, in particular in relation to the harmonization of the procedures'. This report was 'to be accompanied, where appropriate, by new legislative proposals'.

A 'choice opportunity' (Cohen, March and Olsen 1972) had thus appeared which enabled the Commission officials to table proposals for the reform of the regulatory environment. 'Choice opportunities are occasions when an institution is expected to produce behavior that can be called a decision. Opportunities arise regularly and occasions for choice are declared routinely' (March and Olsen 1989, p. 13). Kingdon (1995, p. 201–203) uses the term 'policy window' to explain the same phenomena. 'An open window is an opportunity for advocates to push their pet solutions or to push attention to their special problem'. 'Advocates of a new policy initiative not only take advantage of politically propitious moments but also claim that their proposal is a solution to a pressing problem'. Here, the policy window opened (or the choice opportunity arose) as expected. 'Sometimes, windows open quite predictably. Legislation comes up for renewal on a schedule, for instance, creating opportunities to change, expand, or abolish certain programs'.

9.2 THE 2003 ELECTRONIC COMMUNICATIONS FRAMEWORK

The Commission took advantage of the open policy window to pursue its pet policy solution: a modified institutional and regulatory framework more fully incorporated into the EC's supranational framework. Whereas the 1998 package eliminated PTT monopolies over the sector, the Commission intended the new framework to react to increasing competition by rolling back regulation and relying on competition law as the markets become more competitive. In examining the development of the 2003 electronic communications framework, it is helpful to rely upon a synthetic theoretical framework based on stages or levels of analysis.

Super-systemic or Macro Level of Analysis

The super-systemic or macro level of analysis (Peterson 1995a; see also Peterson and Bomberg 1999; and Peterson 2001) is where the overall development of the 2003 electronic communications framework can be conceptualised. The focus of this level of analysis is the process of integration or Europeanisation of the telecommunications sector and not the day-to-day legislative processes discussed in Parts II and III. At this level, it is apparent that the technological and economic pressures outlined in chapter two drove the process. Meta theories, such as neofunctionalism and intergovernmentalism, are useful in analysing the movement towards establishing a new regulatory framework at the European level.

In line with the neofunctionalist concept of spillover (Haas 1968), and following the implementation of the 1998 package as discussed in previous chapters, the integration process in telecommunications continued to feed upon itself. In other words, the development and implementation of the 1998 package helped to provoke further, related acts of integration in the telecommunications, information technology and broadcasting sectors. As discussed below, social actors, including interest group and industry representatives, also began to lobby for further integration. Following the review of the regulatory framework in November 1999, the Commission took advantage of the open policy window to launch the eEurope initiative in December 1999. This aimed to give fresh impetus to the integration process by promoting 'greater coherence between the policies and paces in the Member States'.[12] The Commission explained the need for additional reform:

> The liberalisation of the market for telecommunications infrastructures and services in the Union took a new direction from the 1st of January 1998. The evidence of falling prices and increased consumer choice indicates that this policy is delivering positive results. Nevertheless, much needs to be done. The distribution of the benefits of competition is still uneven from one Member State to another. Truly pan-European services are still underdeveloped, partly as a result of fairly different and sometimes excessive licensing conditions and procedures.[13]

As with earlier reform efforts, exogenous pressure from technological and economic developments continued to spur on the integration process. The Commission explained that:

> eEurope is a political initiative to ensure the European Union fully benefits for generations to come from the changes the Information Society is bringing. These changes, the most significant since the Industrial Revolution, are far-reaching and global . . . What is emerging is often referred to as the *new economy*. It has tremendous potential for growth, employment and inclusion. Yet Europe is not fully exploiting this potential as it is not moving fast enough into the digital age.[14]

The eEurope initiative aimed to accelerate Europe's movement into the digital age and established, as its primary objective, to ensure that all Europeans have access to the internet. 'The objective of the eEurope is ambitious. It aims to bring everyone in Europe – every citizen, every school, every company – online as quickly as possible.'[15] 'The ultimate goal', explained Information Society Commissioner Erkki Liikanen, 'was to grab the opportunities of the Internet for competitiveness, growth, employment and cohesion'.[16]

Highlighting a core assumption of liberal intergovernmentalism, that the Member States are rational self-interested actors that use the EU as a forum to manage their complex interdependence (Putnam 1988; Moravcsik 1994), the Lisbon European Council in March 2000 recognised the need for a coordinated approach and established the ambitious objective for Europe to become the 'most competitive and dynamic knowledge-based economy in the world by 2010'. The Heads of the Member States identified an urgent need for Europe to exploit the opportunities presented by the 'new economy' and, particularly, the internet. To achieve these objectives, the Heads of the Member States called for a regulatory framework that would be pro-competitive and responsive to convergence and the changing marketing and technological conditions, as well as reinforce competition and take into account the increasing speed of developments in the sector.[17] At the European Council meeting in Stockholm, Sweden, in March 2001, the Heads of States set the ambitious goal of adopting a new regulatory framework by the end of 2001. The Conclusions of the Presidency stated: 'the telecommunications package should be adopted as soon as possible this year in order to offer the sector a level playing field in which rules are applied in a harmonised manner across the Union' (*RAPID*, DOC/01/06, 26 March 2001).

In line with intergovernmentalism, the Member States had greater control than other actors in creating the electronic communications framework due to their institutional authority for positive integration (Scharpf 1999). Notwithstanding the Commission's success in coordinating the development of the new framework, and as discussed further below, the Member States once again were able to impose limits on the Commission's attempt to Europeanise regulatory authority over the sector.[18]

Sub-systemic Level of Analysis

The sub-systemic level of analysis is where policy-shaping decisions are taken as policy options are formulated (Peterson and Bomberg 1999). Chapter four examined the sub-systemic level, in a temporal sense, as the policy formulation stage until the Commission forwards a draft proposal to the Council of Ministers and the European Parliament. Continuing to take advantage of the open policy window, and in response to the perceived exogenous challenges posed by the 'new economy', in April 2000 the Commission put forth draft proposals for directives on: (i) a common regulatory framework for electronic communications networks and services (Framework Directive); (ii) the authorisation of electronic communications networks and services; (iii) access and interconnection; (iv) universal services; and (v) data protection. In line with the pluralistic policy formulation process examined in chapter four, the Commission explained that these proposals were the result of extensive consultation with affected interests:

> These proposals are the outcome of extensive public consultation, including the Green Paper on the convergence of the telecommunications, media and information technology sectors ... as well as the 1999 Communications Review of the existing regulatory framework and the public consultation that followed.[19]

The Commission's initial proposals were thus based largely on expert reports and consultations, including the 1999 Review discussed in section

9.1. The proposals were also subject to further consultations, including a public hearing in May 2000 that drew several hundred industry experts, representatives from interest groups and other interested parties.

The policy network expanded as membership spilled over from telecommunications to new industries that emerged as a result of new technologies (for example, the internet), and as more groups became affected by the broadening of the regulatory framework (for example, marketing companies that rely on new technologies). The electronic communications policy network is thus increasingly characteristic of an 'issue network' where 'membership is fluid and non hierarchical, the network is easily permeated by external influences, and actors are highly self-reliant' (Peterson 1995a). Indeed, more than 120 actors and organisations representing a wide array of industries commented on the commissions proposals, including:

- LINX and AOL Europe (internet service providers);
- ITV and BBC (British television stations);
- EACEM (representing consumer electronic manufacturers);
- EADP (representing directory and database publishers);
- European Disability Forum (representing disabled EU citizens);
- European Broadcasting Union;
- World Forum for Digital Audio Broadcasting;
- Public Utilities Access Forum;
- Independent Television Commission;
- Federation of European Direct Marketing;
- the Japanese Government;
- Corning (technology manufacturing corporation); and
- US Council for International Business;

Non-sectoral influences with weak resource dependencies (Peterson 1995a) thus appear to have gained access to the electronic communications policy network. In addition, longstanding members of the EU telecommunications policy network, including OFTEL, British Telecom, France Télécom, Deutsche Telekom, ETNO, INTUG and the European Telecommunications Platform, submitted comments on the Commission's proposals.

Highlighting the important effect these early consultations can have on policy development, Commissioner Liikanen cited the thorough consultation with interested parties and impact analysis as the reason the proposals subsequently moved rapidly through the inter-institutional bargaining stage of the policy-making process.

Apart from preparatory documents that were widely put to public consultation, the draft legislation was also put on the Web in the form of working documents. The time used for this exercise proved well spent in the subsequent decision-making process, since the legislation was adopted by the Council and the European Parliament in only [seventeen] months.[20]

Systemic Level of Analysis

At the systemic level of analysis, bargaining is primarily inter-institutional and new institutionalism is the 'best' theory for explaining policy development (Peterson and Bomberg, 1999). Chapter five examined the systemic level, in a temporal sense, as the inter-institutional bargaining stage beginning when the European Parliament and Council receive a draft proposal from the Commission. The inter-institutional bargaining stage for the proposals began in July 2000, when the Commission forwarded them to the EP and the Council. At the same time, the Commission warned the institutions of the need to adopt the proposals rapidly:

> [U]nder the traditional legislative procedures the proposals would take up to three years to be fully implemented in the Member States. In such a dynamic market, this is too long. Therefore, the Council and the European Parliament are invited to make all possible efforts to accelerate the legislative process. Moreover, Member States can through their own decisions speed up liberalisation and urgently address issues that would give consumers more choice and lower prices for high-speed access to the Internet.[21]

Under the proposed framework, the NRAs were to play a large role in the implementation of the directives. Whereas the Commission had previously hoped to establish an ERA, the new proposal instead sought to set up a transnational committee of NRAs to coordinate implementation at the EU level. Thus, in response to exogenous technological pressures and the 'policy window' discussed above in section 9.1, the proposal aimed to create a new regulatory environment. The proposed framework would give the NRAs authority to make decisions based on a three-part analysis of: 1) the definition of relevant markets; 2) the identification of dominant companies; and 3) the formulation of remedies. More specifically,

1. The Commission would adopt Recommendations periodically that identify, consistent with principles of competition law, the markets in the electronic communications sector that may have characteristics that would justify imposing regulatory obligations set out in specific directives.[22] Taking 'utmost' account of the Commission Recommendations and guidelines on market analysis, it would be up to the NRAs to define the

relevant markets that are appropriate to particular national circumstances and relevant geographic markets within their Member State.

2. The NRAs would then analyse the relevant markets in their territories to determine whether they are competitive and whether one or more companies maintain dominance – hold significant market power (SMP) – in the market. An undertaking is deemed to have SMP 'if either individually or jointly with others, it enjoys a position equivalent to dominance, that is to say a position of economic strength affording it the power to behave to an appreciable extent independently of competitors customers and ultimately consumers.'[23]

3. If an NRA determines that one or more companies has SMP in that market, the NRA must impose specific obligations from the menu of remedies proposed by the directives.[24]

According to the proposed framework, each of these tasks was to be undertaken by the NRAs. However, the draft framework also proposed to give the Commission authority to require the NRAs to amend or withdraw proposals that it felt were inconsistent with Community law.[25] With this veto power, the Commission hoped to balance out the increased NRA authority in the new regulatory framework and hopefully ensure harmonisation. As chapter five found, and in line with rational choice institutionalism, institutions frequently seek to defend or expand their own powers and the bargaining over institutional design can lead to delays in the adoption of important telecommunications legislation. As the discussion below indicates, the Commission's authority to overrule NRA decisions was one of the most controversial elements of the proposed regulatory framework and threatened to delay its adoption.

As with several of the Commission's proposals discussed in chapter 6, the proposal for veto authority sought to extend supranational competencies beyond what the Member States were willing to accept. Certain Member States, particularly Austria, Germany and Spain, opposed the Commission's proposed veto authority, in large part because they wanted to protect their national sovereignty – they felt that their NRAs were better placed to assess the market because they were closer to it and could make better decisions within their particular national regulatory systems.[26] As with the discussion of the ERA in chapter 6, these Member States also wanted to limit the shift of regulatory authority to the European level.[27] These Member States may also have resisted further transfer of authority to the EU level partly because of their stake in their national telecommunications operators and lingering governmental influence over the NRAs.[28] Indeed, the different national institutional arrangements, perspectives and ideologies discussed in chapter two undoubtedly influenced, and continue to influence, Member State positions at the EU level.

In April 2001, the Swedish Presidency of the Council put forth a compromise to weaken the Commission's authority. Under the compromise, the Commission would be allowed to publish 'a detailed opinion' expressing its concern that an NRA's proposed measure is incompatible with Community law, but would not have the opportunity to compel the NRA to alter or withdraw a planned measure (*European Report*, 28 March 2001). As indicated in chapter five, in order to modify the Commission's proposal (at the Second Reading stage), and thus avoid a conciliation committee with the EP, the Member States would need unanimous agreement. According to rational choice institutionalism, these formal policy-making rules shape the rational actions of political actors (Tsebelis 1990), such as the Member State representatives. Accordingly, after ten hours of discussions, the EU Telecommunications Ministers unanimously rejected the Commission's proposed injunctive powers and instead approved the Swedish compromise. 'Obviously today there's a great difference with selling toothpaste and cars, where there are normal competition rules', said the Swedish Minister chairing the meeting, Bjorn Rosengren. 'There is [also] quite a strong national feeling for national regulatory authorities' (*The Financial Times*, 6 April 2001). Acknowledging the power of the Member States to block Commission proposals that they do not collectively agree with (Scharpf 1988), Commissioner Liikanen indicated that he was not surprised by the Council decision, that the time had not yet come when the Commission would be given the right to 'supervise' telecommunications, although he was confident the time would come (*European Report*, 7 April 2001).

Interest group pressure, however, continued to mount as industry representatives increasingly lobbied EU policy makers for reform. Mobile phone operators, in particular, lobbied in favour of the proposal to give the Commission veto powers because they felt it would lead to increased harmonisation (*The Financial Times*, 6 April 2001). UNICE, the European employers' organisation, issued a press release calling for the Council to support the Commission proposal because it feared that without such arrangements, 'which is a fundamental guarantee of coherence at the EU level', there would be no single market in telecommunications (*European Report*, 5 December 2001). In October 2001, the European Telecommunications Network Operators (ETNO) met with Rik Daems, Telecommunications Minister from Belgium, which had the Presidency of the Council, to express its support for Commission authority:

> ETNO asked for light regulation to be applied uniformly throughout the Member States. It favours giving the Commission real power to counterbalance the large

degree of autonomy enjoyed by the National Regulating Authorities (NRA) ... ETNO has also asked the institutions to bear very much in mind the need to strike the right balance between adopting a regulatory framework quickly and adopting the right framework. (*European Report*, 24 October 2001).

Industry viewed positively the increased powers of the Commission because it favoured a larger EU role to help guard against dissimilarities in Member State implementation (EUObserver.com, 13 December 2001). The Commission also refused to compromise its position:

> The Commission ... was not willing to accept the loss of the veto it would like over national regulators' decisions ... Member states argued that instead the Commission should have the power to 'name and shame' national regulators' decisions with which it did not agree. Emboldened by strong support from the European Parliament, Mr. Liikanen refused to yield his ground. (*The Financial Times*, 6 April 2001)

Finally, and perhaps most importantly, the European Parliament used its increased agenda-setting authority in this area (Tsebelis 1994) to bolster the Commission position. In October 2001, Eino Paasilinna, the EP's *rapporteur*, complained that, by rejecting the Commission's veto authority, 'the Council was undermining the EU telecommunications market and paving the way for a fragmented system' (*European Report*, 13 October 2001). Consistent with chapter five's finding that only a small fraction of MEPs belong to the epistemic community with expert knowledge in telecommunications (Haas 1992, p. 3), and that these knowledgeable MEPs have significant influence over EP decision making, the views of rapporteur Paasilinna were upheld by MEPs from other political groups (*European Report*, 13 October 2001). In its second reading, the EP's Committees on Legal Affairs and Industry thus reiterated the EP's first reading position of empowering the Commission with veto authority (*European Report*, 1 December 2001). Highlighting its increased confidence, the European Parliament indicated that, if the Council changed its mind on this issue, it would be willing to make a trade-off with the Council by accommodating other amendments. Highlighting its increased influence, the EP subsequently agreed to approve the proposals only after the Member States agreed informally to grant the Commission veto authority (EUObserver.com, December 13, 2001).

> In contrast to the Council's first reading in April 2001, which was over the Framework Directive individually, the Council's second reading in December 2001 was over the whole package, so that opened a lot of opportunity for trade-offs between the various directives. By reviewing the Framework Directive within the context of the whole package of directives, and after a lot of discussion and give and take, a global balance was achieved.[29]

In December 2001, facing pressure from industry and the European Parliament, the Member States agreed to a compromise. The Member States agreed to accept Commission authority to overrule NRA decisions with regard to the definition of relevant markets and the identification of dominant companies, but not for the selection of remedies.[30] Rik Daems, Belgian Telecommunications Minister, was instrumental in getting the Member States to agree informally to the compromise (*Financial Times*, 13 December 2001). In urging the European Parliament to vote for the package, Commissioner Liikanen explained that all of the institutions had compromised:

> The package of compromise amendments before you today has a broad level of support in the Council. They are of course a compromise. Parliament doesn't get everything it wants here. Nor does Council. (Nor does the Commission for that matter). But the Commission nonetheless believes that it is a balanced compromise. And my assessment is that neither institution is likely to get more from conciliation. (*RAPID*, 11 December 2001)

The European Parliament subsequently voted to accept the compromise deal submitted by the Belgian Presidency. Commissioner Liikanen commented:

> I am delighted by today's decision by the European Parliament. The Member States and the European Parliament have delivered on all key elements of the Lisbon agenda dealing with Information Society by the end of this year. This agreement is a major boost for Europe's future economic growth and employment. (*RAPID*, IP/011801, 12 December 2001)

Despite the difficult political compromise over the Commission's veto authority, there was widespread agreement on the need to adopt the directives rapidly to create a stable regulatory environment. Heeding the European Council's call for adoption before the end of 2001, the directives were informally agreed upon in December 2001 and formally adopted by the Council in February 2002. From July 2003, the draft directives were applied.[31] Commissioner Liikanen echoed how remarkable it was for the regulatory framework for electronic communications to have 'been completed in precisely 17 months. This represents a remarkable achievement for a package of this size and complexity. And it shows how rapidly the European institutions can achieve clear objectives when there is widespread support.'[32]

The development of the new framework also highlights the European Parliament's increased decision-making authority (Maurer 2003). Several Member State representatives indicate that the Council would probably not have agreed to give the Commission veto authority over certain NRA decisions if the European Parliament had not gained increased authority under the co-decision procedure.[33] Indeed, the Member States recognised the importance of working with the European Parliament to find a resolution to the dispute.[34] In line with the joint-decision trap (Scharpf 1988), however, the Member States were nevertheless once again able to impose limits on the transfer of authority to the EU by refusing to give the Commission explicit power to overrule NRA decisions with regard to the selection of remedies.

Nevertheless, highlighting the extent to which the sector has been Europeanised, the new framework covers all electronic networks and services, including fixed and mobile telecommunications networks, networks for terrestrial broadcasting and satellite and internet networks, whether used for voice, data, fax or image transmission. The electronic communications framework consolidated and replaced the directives and decisions that comprised the ONP regulatory framework (see Figure 9.1).

The electronic communications framework implemented in 2003 is simplified and adapted to the convergence of the telecommunications, information technology and audiovisual sectors. Whereas the ONP regulatory framework was focused on the creation of a competitive market and the rights of new entrants, the electronic communications framework is designed to manage the challenges of liberalised markets, including a lighter regulatory touch where markets have become more competitive. As this book has repeatedly demonstrated, exogenous pressures, including technological and economic developments, have generally driven change in European telecommunications. In anticipation of continuing technological change, the framework is technology-neutral, so that it can encompass new, dynamic, more developed and unpredictable markets, with more market players, although content services still remain outside its scope.

Also as part of the electronic communications framework, and following another Commission proposal, the Council and Parliament adopted a regulation on the unbundling of the local loop in December 2000.[35] The institutions had identified, through 'feedback' (Kingdon 1995), that there was a problematic condition with prior liberalisation efforts – the former PTTs still largely dominated local communications. The institutions also intended the regulation to drive down tariffs and encourage innovation in internet access, thus furthering the eEurope objective of moving Europe into the digital age. In addition, the Commission used its authority under Article 86 of the EC Treaty to adopt a sixth directive on competition in the markets for electronic communications networks and services in September 2002.[36] This Liberalisation Directive replaces and consolidates the directives discussed in Part III of this book (see Figure 9.2).

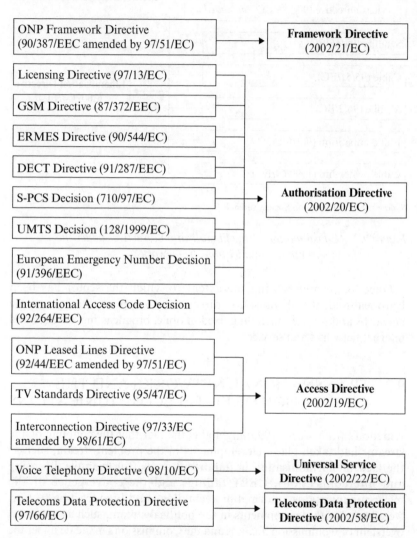

ONP Regulatory Framework
In force January 1998–July 2003

Electronic Communications Framework
In force since July 2003

Article 95 Directives/Decisions

ONP Framework Directive
(90/387/EEC amended by 97/51/EC)
→ **Framework Directive**
(2002/21/EC)

Licensing Directive (97/13/EC)

GSM Directive (87/372/EEC)

ERMES Directive (90/544/EC)

DECT Directive (91/287/EEC)

S-PCS Decision (710/97/EC)
→ **Authorisation Directive**
(2002/20/EC)

UMTS Decision (128/1999/EC)

European Emergency Number Decision
(91/396/EEC)

International Access Code Decision
(92/264/EEC)

ONP Leased Lines Directive
(92/44/EEC amended by 97/51/EC)

TV Standards Directive (95/47/EC)
→ **Access Directive**
(2002/19/EC)

Interconnection Directive (97/33/EC
amended by 98/61/EC)

Voice Telephony Directive (98/10/EC)
→ **Universal Service
Directive** (2002/22/EC)

Telecoms Data Protection Directive
(97/66/EC)
→ **Telecoms Data Protection
Directive** (2002/58/EC)

Source: DG Information Society official

*Figure 9.1   Consolidation of the ONP directives under the electronic
communications framework*

Prior Regulatory Framework   2003 Electronic Communications Framework

Artice 86 Directives

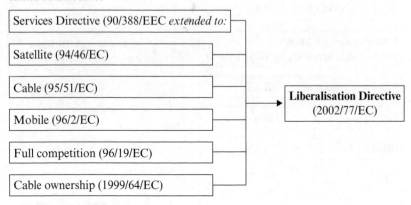

Source: DG Information Society official

Figure 9.2   Consolidation of the liberalisation directives under the electronic communications framework

Once again highlighting the extent to which the sector has been Europeanised, the liberalisation directive also applies to all electronic networks and services, including fixed, mobile, broadcasting, satellite and internet networks and services.

9.3   TRANSNATIONAL NETWORKS AND THE REGULATION OF EU TELECOMMUNICATIONS

Kingdon's framework (1995) highlights the fact that the coupling of the streams had taken place. Developments in the problem stream, such as the lack of harmonisation in important areas (for example, licensing and interconnection), CEPT failings and the convergence of the telecommunications, information technology and broadcasting sectors, were linked with developments in the political stream, such as increased oversight of Commission finances and the Commission's mandate under the ONP framework to undertake a review and propose further measures. Both of these streams were linked with the policy stream and the Commission's pet solution: a modified institutional and regulatory framework, adapted to the convergence of technologies and more thoroughly incorporated into the supranational EC.

Although the epistemic community discussed above in section 9.1 had determined that the 'political climate' did not allow the Commission to establish a regulatory authority in telecommunications at the European level (Watkinson and Guettler 1999), the Commission nevertheless was able to create several committees under the electronic communications framework to help the NRAs coordinate their regulatory activities at the EC level. Eberlein and Grande (2005, pp. 99–100) explain that transnational networks such as these arise as agents of informal harmonization because,

> [d]espite the rising need for uniform European rules, the European level still lacks the formal powers and the institutional capacities needed to establish the appropriate rules and to monitor and enforce their compliance and transposition in the member states … Since the solution that most suggests itself for the problem from a functional perspective – namely, greater centralization of formal powers – has thus far been barred politically, a *regulatory lacuna* has emerged that threatens to considerably weaken the effectiveness of the 'regulatory state' in Europe.

Eberlein and Grande (2005, p. 100) conclude that, due to the deficiency of regulatory authority at the European level, alternative routes have been necessary to ensure that regulatory activities are sufficiently Europeanised: the 'existing regulatory gap has been closed by new types of informal institutions, the *transnational regulatory networks*'.

In line with the thesis put forward by Eberlein and Grande, and in the light of the Commission's failure to centralise formal powers in a European regulator (see chapter 6), several committees have been formed at the European level to facilitate coordination between the NRAs. A European Regulators Group for Electronic Communications Networks and Services was established within the Community framework in order to facilitate cooperation between the NRAs and the Commission.[37] The European Regulators Group is composed of NRAs and was designed:

> to provide an interface for advising and assisting the Commission in the electronic communications field. The Group should also provide an interface between national regulatory authorities and the Commission in such a way as to contribute to the development of the internal market. It should also allow cooperation between national regulatory authorities and the Commission in a transparent manner so as to ensure the consistent application in all Member States of the regulatory framework for electronic communications networks and services. The Group should serve as a body for reflection, debate and advice for the Commission in the electronic communications field . . . Close cooperation should be maintained between the Group and the Communications Committee established under the Framework Directive. The work of the Group should not interfere with the work of the Committee.[38]

By further integrating the NRAs into the Community structure through the establishment of the European Regulators Group, the Commission hoped to achieve a more consistent application of the electronic communications framework across the Member States than had been achieved under the ONP regulatory framework. Information Society Commissioner Erkki Liikanen explained that, through the European Regulators Group, the NRAs are 'first and foremost responsible for guiding the application of the new framework'.[39] Its work is focused 'on the tasks linked to day-to-day supervision of implementation of the new regulatory framework for electronic communications networks and services'.[40] The role of the European Regulators Group was broadened in September 2004 to include advising and assisting the Commission 'on any matter related to electronic communications networks and services within its competence, either at its own initiative or at the Commission's request'.[41]

The European Regulators Group has already made significant strides toward facilitating the harmonised and coherent application of the electronic regulatory framework. For example, in April 2004, it adopted a *Common Position on the Approach to Appropriate Remedies in the New Regulatory Framework*.[42] Information Society Commissioner Erkki Liikanen commented, 'This is an important milestone for the new regulatory framework ... It provides an important step towards delivering the full potential of our reforms through greater predictability, coherence and a more harmonised approach to the way markets operate across the Union'.[43] Competition Commissioner Mario Monti added, 'I very much welcome the principles which this document establishes. In particular, I believe that the Common Position strikes the right balance between competition based on alternative infrastructure and competition based on access to existing infrastructure held by operators who have significant market power'. [44]

The electronic communications framework also created other new committees to assist the Commission and facilitate a harmonised implementation of the framework, including the Communications Committee discussed in chapter four (replacing the Open Network Provision and the Licensing Committee), the Radio Spectrum Committee and the Radio Spectrum Policy Group. These committees provide further platforms for the Commission and NRAs to exchange information on regulatory activities. Another group, the Independent Regulators Group, was established outside of the EU framework in 1997 'to share experiences and points of views among its members on issues of common interest such as interconnection, prices, universal service, and other important issues relating to the regulation and development of the European telecommunications market'.[45] With 31 members (the EU Member States plus Bulgaria, Romania, Iceland, Liechtenstein, Norway and Switzerland), the Independent Regulators

Group has 'developed into quite an important talking shop, where best practices can be exchanged among regulators'.[46]

These transnational networks of NRAs have gone a long way towards meeting the need for regulatory capabilities at the European level – the need having resulted from the increased harmonisation of telecommunications policies at the European level and the corresponding failure to establish an ERA to oversee implementation at the national level. As Eberlein and Grande (2005, p. 103) conclude,

[I]t seems that both the formation of transnational networks promoted at EU level and the intergovernmental networking among national regulatory authorities make important contributions towards at least blunting the dilemma … [that arises due to] the functionally necessary harmonization and the politically barred centralization.

In addition to further institutionalising the role that the NRAs play in coordinating the implementation of the directives, the new framework also seeks to overcome some of the other deficiencies of the ONP framework. In particular, the electronic communications framework largely eliminated the granting of individual licences and instead, requires Member States to establish general authorisations for all types of electronic communications services and networks, including fixed and mobile networks and services, voice and data services, and broadcasting transmission networks and services.[47] As Commissioner Liikanen explained,

the new framework drastically cuts away the unnecessary red tape which obstructs entry to national markets by replacing individual licences by general authorisations to provide services. This removes any possibility for regulators to insist, as many do today, on checking compliance with licence conditions before allowing telecoms operators to begin to offer services to consumers and businesses.[48]

Thus, after several frustrating years and unsuccessful attempts at licensing harmonisation (see chapter six), the electronic communications framework establishes a system whereby the Member States can no longer use the instrument of individual licences to regulate the sector. NRAs cannot require service providers to obtain explicit administrative decisions before providing services.[49] Finally, and crucially, the CEPT no longer plays a role in the EC-level licensing regime.[50]

Accordingly, the Commission once again has been able to take advantage of an open policy window to further the integration process. Several committees have been established to facilitate the implementation of the new regulatory framework by the NRAs. Nevertheless, in line with Scharpf's joint-decision trap (1988), and highlighting the greater control that national government representatives have in policy development and in creating and modifying institutions, the Member States were able to resist the Commission's effort to gain veto authority over all NRA decisions concerning the implementation of the 2003 electronic communications framework.

9.4   CONCLUSION

The implementation of the ONP legislation examined in previous chapters led to a new phase of policy development. John Kingdon's model (1995) is useful in highlighting the separate streams of problems, politics and policies that characterise the early stage of policy making. As with prior iterations of the policy-making process examined in previous chapters, the Commission took advantage of an open policy window to garner support for its favoured policy line: a new regulatory framework more firmly anchored within the EC.

In examining the development of the 2003 electronic communications framework, a synthetic approach helps to elucidate the multi-level character of the policy-making process. In particular, meta-theories help to shed light on the overall development of the new framework at the super-systemic level, while the expansion of the policy network is visible at the sub-systemic level and inter-institutional bargaining is prevalent at the systemic level. The import of the new framework is also recognised, as it furthers the integration process through the regulation of all electronic communications markets, eliminates the role that the CEPT had played with regards to licensing harmonisation and establishes an EC-wide licensing regime. The new framework nevertheless still falls short of some of the Commission's earlier ambitions, such as the establishment of a European Regulatory Authority. Thus, although the Commission has continued to be a successful policy entrepreneur in telecommunications, it has indeed faced limits in its ability to Europeanise the sector. In particular, the Commission's success in the area of positive integration has been dependent on its institutional authority and the support of the Member States.

Because the Commission has been unable to establish an ERA, the European Regulators Group composed of NRAs was formed to facilitate implementation of the new framework. Although the Commission, with EP support, gained authority to oversee some NRA decisions concerning the implementation of the framework, the Member States refused to give the Commission power to overrule NRA decisions on remedies for uncompetitive markets. Since the Member States continue to resist the Commission's efforts to shift significant regulatory powers to the EU level, and because of their privileged position in the policy-making process, it is unlikely that a regulatory agency will be established at the European level in the foreseeable future. As such, transnational networks of national regulatory bodies, such as the recently established European Regulators Group and the Communications Committee, will undoubtedly remain important forums for coordinating EU-wide implementation.

NOTES

1. Enterprise and Information Society Commissioner Erkki Liikanen speech at the 'Information Technology and Telecommunications in the EU 2002–2005: From Visions To Solutions' conference, Estonia Paernu, 20 September 2002.
2. Commission Press Release, 'Commission adopts 7th Report on liberalisation of telecommunications in the Member States', IP 01/1679, 28 November 2001.
3. Interview, INTUG representative, telephone, June 2000; interview, INTUG representative, telephone, November 2000.
4. Interview, London, May 1999.
5. Interview, DG Information Society (DG XIII) official, Brussels, February 1998.
6. In the political stream, consensus is built by bargaining based on self-interested behaviour, more than through persuasion (Kingdon 1995, p. 199; see also Scharpf 1988).
7. Interview, ETO official, telephone, July 2000.
8. Interview, DG Information Society (DG XIII) official, Brussels, February 1998, July 2000.
9. Interview, DG Information Society (DG XIII) official, Brussels, July 1999.
10. Robert Verrue speech 'Next challenges facing the 1998 European Telecommunications Regulatory Framework', EUI Institute, Florence, 13–14 November 1998.
11. Interview, DG Information Society (DG XIII) official, Brussels, July 1999
12. Information Society Commissioner Erkki Liikanen speech at the 'Accelerating the pace towards Stockholm and beyond', British Embassy Seminar, Brussels, 22 February 2001.
13. 'eEurope: an Information Society for all,' Communication on a Commission Initiative for the Special European Council of Lisbon, 23–24 March 2000.
14. *Ibid.*
15. *Ibid.*
16. Information Society Commissioner Erkki Liikanen speech at the 'Accelerating the pace towards Stockholm and beyond', British Embassy Seminar, Brussels, 22 February 2001.
17. See 'eEurope 2002: an Information Society for all', prepared by the Council and the European Commission for the Feira European Council, 19–20 June 2000.
18. See chapter six for a discussion of the Member States' institutional authority for positive integration and their ability to prevent the establishment of an ERA and the implementation of a coherent EC-level licensing regime.
19. European Commission, Proposal for a Directive of the European Parliament and of the Council on a Common Regulatory Framework for Electronic Communications Networks and Services, 2000/0184 (COD), COM(2000) 393 final, Brussels, 12.7.2000.
20. Commissioner Liikanen speech at the 'Better Regulation' Informal Council (Competitiveness), Nyborg, 12 October 2002, RAPID Speech 02/477.
21. See 'eEurope : an Information Society for all', Communication on a Commission Initiative for the Special European Council of Lisbon, 23–24 March 2000.
22. See, for example, Commission Recommendation of 11 February 2003 on relevant product and service markets within the electronic communications sector susceptible to *ex ante* regulation in accordance with Directive 2002/21/EC of the European Parliament and of

the Council on a common regulatory framework for electronic communications networks and services, OJ 8.5.2003 L 114/45.

23. Article 14(2) of Directive 2002/21/EC (Framework Directive), OJ L 108, 24.4.2002. This provision is taken from the definition of dominance put forward by the European Court of Justice in its competition case law. See Case 27/76, *United Brands*, (1978) ECR 207.

24. Article 16(4) of Directive 2002/21/EC (Framework Directive), OJ L 108, 24.4.2002. Indeed, the new regulatory framework recognised that telecommunications regulation deals primarily with responding to SMP. Under the new framework, the obligations for companies determined to have SMP include transparency, non-discrimination, accounting separation, access to and use of specific network facilities (including interconnection), price controls and cost accounting, making leased lines available, and carrier selection and pre-selection. See Articles 9–13 of Directive 2002/19/EC (Access Directive), OJ L 108, 24.4.2002; and Articles 18–19 of Directive 2002/22/EC (Universal Service Directive), OJ L 108, 24.4.2002.

25. European Commission, Proposal for a Directive of the European Parliament and of the Council on a Common Regulatory Framework for Electronic Communications Networks and Services, Article 6(4) of Proposal 2000/0184, COM(2000) 393 final, 12.7.2000.

26. Interview, Adviser to German Economic Ministry, telephone, June 2005; interview, Adviser to Austrian Permanent Representation, telephone, June 2005.

27. *Ibid.*

28. Interview, Adviser to British Permanent Representation, telephone, June 2005. The British Adviser explained that certain Member States, such as France and Germany, still had significant stakes in their national telecommunications operators and maintained some influence over their NRAs. Because of this, these governments tended to have greater 'unease about transferring authority to the EU level'. Prior privatisation and liberalisation in the UK, combined with the early establishment of an entirely independent regulator, OFTEL, was a factor in the UK's support of EU-level regulatory authority.

29. Interview, Adviser to Belgian Permanent Representation, telephone, May 2005.

30. The Commission was also satisfied with the compromise position and explained that it 'represents a good balance between the two positions, concentrating the Commission's power to intervene on the two issues central to maintaining consistency of regulatory action'. Commission Opinion, COM(2002) 78 final, 7 February 2002.

31. Certain transitional measures remained in force until the new institutions and procedures of the new framework were ready to replace them. Also, the Data Protection Directive was not applied until October 2003. Moreover, there were a number of Member States that failed to transpose the new regulatory framework into national law by the July 2003 deadline. In April 2004, the Commission took six Member States (Belgium, Germany, Greece, France, Luxembourg and the Netherlands), to the European Court of Justice for failing to implement fully the new regulatory framework.

32. Commissioner Erkki Liikanen speech, 'The EU Communications Regulatory Framework', Forum for EU-US Legal-Economic Affairs, Brussels, 3 April 2002.

33. Interview, Adviser to Belgian Permanent Representation, telephone, May 2005; Interview, Adviser to Swedish Permanent Representation, telephone, May 2005.

34. Following their October 2001 meeting, the telecommunications ministers pressed their permanent representatives to continue talks with MEPs in an attempt to avoid going through conciliation (*European Report*, 17 October 2001).

35. Regulation (EC) No 2887/2000 of the European Parliament and of the Council of 18 December 2000 on Unbundled Access to the Local Loop, L336/4, 30 December 2000.

36. Commission Directive 2002/77/EC of 16 September 2002 on Competition in the Markets for Electronic Communications Networks and Services, L249/21, 17 September 2002.

37. Commission Decision of 29 July 2002 establishing the European Regulators Group for Electronic Communications Networks and Services, L200/38, 30 June 2002.

38. *Ibid.*

39. Commissioner Erkki Liikanen speech at the inaugural meeting of the European Regulators Group in Brussels, 25 October 2002.

40. Commission Decision of 14 September 2004 amending Decision 2002/627/EC establishing the European Regulators Group for Electronic Communications Networks and Services, L293/30, 16 September 2004.
41. ERG Common Position on the approach to Appropriate Remedies in the New Regulatory Framework, ERG Common Position, 1 April 2004, ERG (03) 30 rev 1. Available at: http://www.erg.eu.int/doc/whatsnew/erg_0330rev1_remedies_common_position.pdf.
42. ERG(03)30rev1, available at http://www.erg.eu.int (last visited May 15, 2005).
43. Commission Press Release, 'Commission welcomes agreement among European regulators on competition remedies to be used in field of electronic communications', Brussels, 23 April 2004, IP/04/528.
44. *Ibid*.
45. http://irgis.anacom.pt/site/en/irg.asp (last visited May 14, 2005)
46. Interview, OFTEL representative, London, May 2000.
47. Individual regulatory decisions remain necessary for rights to use radio spectrum, rights to use numbers and rights of way, and obligations related to universal service and significant market power.
48. Commissioner Erkki Liikanen speech, 'The EU Communications Regulatory Framework', Forum for EU-US Legal-Economic Affairs, Brussels, 3 April 2002.
49. Although NRAs are entitled to receive notification so that they can keep an updated register of service providers, service providers do not have to wait for the NRA's reply to this notification.
50. Interview, DG Information Society official, telephone, September 2004.

10. Conclusion: towards a synthetic approach for analysis

> The basic concept in the social sciences should be that of a mechanism rather than of a theory ... [T]he social sciences are light years away from (being able) ... to formulate general-law-like regularities about human behavior. Instead, we should concentrate on specifying small and medium-sized mechanisms for human action and interaction – plausible, frequently observed ways in which things happen. (Ester 1989, p. viii)

This book evaluated the utility of new institutionalism for analysing the EU policy-making process. The model was useful for its recognition of the important role that institutions play in structuring the access of socioeconomic and political actors to the policy-making process. It helped to explain the development of the policy environment at the European level, where supranational institutions, particularly the European Commission, with the support of the European Court of Justice, played a key role in advancing ideas, mobilising interests and promoting supranational solutions to the common challenges faced by the Member States.

New institutionalism was not, however, able to account for many of the important policy developments in the sector. Although each chapter employed new institutionalism as an explanatory mechanism, it was not fully capable of elucidating all of the significant details in any of the 'snapshots'. Despite, for example, new institutionalism's ability to account for the path dependent nature of the PTTs at the national level (and the CEPT at the European level), an explanation of many other phenomena, such as the highly pluralistic policy formulation process and the timing of full liberalisation, required the use of other theoretical models. In sum, a synthetic approach that employs various theoretical models at different levels of analysis was necessary to explain the development of the EU's telecommunications policy.

Along with new institutionalism, the book drew upon policy actor-based approaches and Kingdon's explanatory model of policy change. The purpose of this chapter is to discuss the wider implications of the findings of the book and to explain how it contributes to the development of a theoretical approach for EU policy making. In particular, the final section argues that a synthetic approach is likely to be a valuable – perhaps necessary – tool for understanding and explaining EU policy development in other areas. In the first place, however, the arguments put forward in the individual chapters are summarised and the limitations of the book are acknowledged.

10.1 THE ARGUMENTS ADVANCED IN THE BOOK

This book analysed the characteristics of several policy-making processes and isolated the factors that helped shape policy outcomes in the EU telecommunications sector. Prior to the 1980s, policy making in the British, French and German telecommunications sectors was dominated by closely-knit policy communities, centred around national PTT monopolies (chapter two). During the 1980s and 1990s, when exogenous technological and economic pressures for change in telecommunications began to mount, national institutional arrangements helped to shape reform efforts. In the UK, centralised authority enabled Margaret Thatcher's government to impose liberalisation and widespread institutional reform upon the sector in the early 1980s. The reform efforts of French and German governments, on the other hand, were restrained by trade union opposition and institutionally-entrenched national PTTs. Kingdon's approach helped account for the timing of reform efforts in France and Germany: in France, an event in the political stream (that is, the Right returning to power in 1993) and in Germany, an event in the problems stream (that is, increased financial pressure resulting from German unification), helped to open 'policy windows' which enabled those governments to achieve widespread reform of the telecommunications sector by the mid-1990s.

National institutional arrangements, especially national PTT monopolies, also affected coordination efforts at the European level (chapter three). In 1959, the national PTTs established a European-level regime, the CEPT, to protect their interests and preserve national autonomy over the sector. Because the EC lacked formal authority over telecommunications, the Commission was forced to rely on its informal agenda-setting powers as a policy entrepreneur to institutionalise telecommunications policy at the EC level.

During the late 1970s and early 1980s, the Commission gradually extended its competence in this sector. Even in the absence of an explicit Treaty base legitimising policy leadership in the sector, the Commission was able to mobilise new interests behind a policy of reform. It created and relied upon epistemic communities, established a transnational policy network of affected interests, pursued pluralistic consultations and marshalled widespread support in favour of reform.

Kingdon's framework is again useful to explain the timing of EC policy development. It was not until Etienne Davignon became Industry Commissioner in 1977 (a development in the politics stream), that the Community began to gain legitimacy in telecommunications. Nonetheless, in line with the historical institutionalist claim of path dependency, the CEPT resisted EC encroachment upon its territory and retained significant influence over transnational telecommunications issues during this period. Analysis of the technological and economic pressures that were mounting during this period helped to identify the factors that shifted national government preferences towards an EU-level regulatory framework, leading to the defining moment: the Commission's 1987 Green Paper on telecommunications. The Green Paper put forth a two-pronged strategy of liberalisation and harmonisation, which set the future course of policy development in the sector.

Having, in Part I of the book (chapters one through three), established the theoretical and historical basis for an examination of the EC telecommunications policy-making process, Parts II and III analysed, in detail, two different institutional arrangements for the production of EC telecommunications policies during the late 1980s and 1990s. Part II examined the harmonisation or positive integration of European telecommunications policies through the EC's normal legislative procedures. Part III viewed the application of the EC Treaty competition rules to liberalise or negatively integrate European telecommunications. Each chapter analysed the ways in which the patterns of policy making identified were related to the two institutional settings. The processes of policy formation followed those established by institutional arrangements, in terms of participants and the division of roles between them.

Part II (chapters four through six) analysed the EU's policy-making process for ONP harmonisation legislation, which took place in accordance with consultative mechanisms established by the Commission and the normal legislative procedures set out in the EC Treaty. In line with the path dependent nature of organisational adaptation, the Commission had institutionalised a powerful PTT role in policy formulation through the GAP and SOGT committees until the late 1980s (chapter four). Telecommunications users had limited influence until institutional change in the early 1990s altered the balance of power, a development that helped to foster policy change.

Subsequently, the policy process began to acquire a more clearly defined institutional shape in which Treaty rules and procedures had greater influence on outcomes, firstly because the Commission was able for reasons driven by wider European developments to become a more central and authoritative actor in the sector, and secondly because this new shape appeared to favour a significantly different group of actors (most notably consumers as opposed to telecommunications operators) from the network that had hitherto been dominant in the policy process.

The Commission nurtured a more inclusive interest group system linked to the sector and more responsive consultation committees were established at the European level. As a result, users were able to increase their influence over policy formulation as the policy-making process became increasingly pluralistic. The Commission finally gained a central policy-making role as a policy broker, while two competing advocacy coalitions (centred on either user or national TO interests) sought to influence the early stages of ONP policy formulation.

Successive modifications to the EC Treaty rules have enabled MEPs to become important institutional actors in the inter-institutional bargaining stage of the telecommunications policy-making process (chapter five). As rational actors, Commission officials and Council members have been forced to adapt their strategies and roles in response to these changes in the institutional environment. In line with rational choice institutionalism, Council Members and MEPs have sought to influence the design of EC institutions in order to maximise their own decision-making authority. MEPs attempted to extend their authority in situations where the EC Treaty rules are ill defined and forcefully resisted efforts to curtail their power. On the other hand, Council members sought to limit the transfer of decision-making power to the supranational institutions. Because of the framework nature of the EC Treaty rules, disputes over institutional authority were found to be a salient feature of this latter stage of the ONP policy-making process.

The institutional context of the EC policy-making process has nonetheless enabled the Member States to resist the positive integration of national regulatory frameworks in key areas of the telecommunications sector (chapter six). This is because the harmonisation of European telecommunications policies depends ultimately upon the agreement of national governments in the Council of Ministers. The power relations in existing institutions thus give national representatives a greater degree of collective control in defending common interests (for example, ensuring respect for subsidiarity for actions better handled at the Member State level), than other actors in the development of policies and the creation and modification of institutions. In particular, the Member States have been able to prevent the establishment of a European-level regulatory authority and were able to delay significantly the establishment of an EC-wide licensing regime.

Part II thus uncovered a complex, highly pluralistic and often lengthy policy-making process for the positive integration of European telecommunications policies. Public and private actors were able to influence the speed, scope and substance of European telecommunications harmonisation: Commission officials played a central coordinating role; outside experts and interest groups maintained significant influence over policy formulation; the final adoption of ONP legislation was dependent on the agreement of both MEPs and national government representatives. The Commission's consultation mechanisms and the EC Treaty's normal legislative procedures thus provided an institutional context that enabled many institutional and political actors the opportunity to either enhance or impede policy development.

In contrast to the institutional setting discussed above, Part III (chapters seven and eight) demonstrated that the treaty competition rules provided for a policy-making process that was, at least initially, less pluralistic and substantially less cumbersome. In particular, the Commission maintained exceptional authority under Article 86 of the EC Treaty (formerly Article 90) to issue directives unilaterally. The Commission was thus, in most cases, able to develop and adopt liberalisation directives more easily and more rapidly than harmonisation directives. By the early 1990s, DG Competition officials had taken advantage of their privileged institutional position to issue directives liberalising terminal equipment and value-added services over the opposition of several Member States, including France and Germany (chapter seven). Providing evidence of an institutional bias for furthering integration, the ECJ also repeatedly confirmed the Commission's authority to issue these directives unilaterally. Nonetheless, following the EC's normative shift towards respecting subsidiarity in 1992, the informal agenda-setting authority of the Commission was diminished. The policy formulation process became more pluralistic and the balance of power shifted towards the Member States.

It was not until a 'policy window' (Kingdon, 1995) opened, due to mounting concerns over the development of the European Information Society, that the Commission regained its informal agenda-setting authority (chapter eight). In particular, a shift in French and German preferences towards further liberalisation, and the construction of a 'purpose-built' (Peterson and Bomberg 1999) policy network on the Information Society (the Bangemann group), facilitated Commission efforts. Towards the end of 1994, DG Competition officials began utilising a more authoritative policy style in order to pressure the reticent Member States into full liberalisation; they relied on the Treaty's competition rules to formulate and adopt directives unilaterally liberalising satellite communications, mobile communications and telecommunications infrastructures. If the Commission had been forced to rely on the normal EC legislative procedures, liberalisation would have been delayed. Due in part to the differences in institutionalised power relations, the Commission had comparatively greater success during the 1990s at coordinating the liberalisation (Part III), as opposed to the harmonisation (Part II), of EC telecommunications. This result confirms that institutions are important factors in shaping policy outcomes.

Part IV discusses the development of the 2003 electronic communications framework (chapter nine) and presents the conclusions of the book (chapter ten). Chapter nine relies on Kingdon's approach (1995) to highlight the Commission's ability to take advantage of a 'policy window' to put forth a modified institutional and regulatory framework to increase EC-wide liberalisation and harmonisation, in areas such as the licensing of telecommunications operators, where previous efforts had failed. The 2003 regulatory framework is also adapted to the convergence of the telecommunications, information technology and audiovisual sectors, and is more fully incorporated into the EC's supranational framework. Chapter nine also discusses the Commission's development of the European Regulators Group to meet the need resulting from the increased harmonisation of telecommunications policies at the European level and the failure to establish an ERA to oversee implementation. This chapter, chapter ten, concludes the book with a discussion of the wider implications of the book's findings and how it contributes to the development of a theoretical approach for EU policy making. What follows next, however, is a brief discussion of the book's limitations.

10.2   THE LIMITS OF THE ARGUMENTS ADVANCED

It is important to acknowledge the limits of the arguments advanced in this book. First of all, the claim of an independent Commission initiative in telecommunications cannot be stretched too far. Reform of some kind would probably have taken place in most Member States without Commission action. During the 1980s, the telecommunications sector in all countries was faced with extreme technological and economic pressures. Every Member State had thus begun to review the sector in preparation for regulatory adjustment prior to any major EC initiative. Moreover, as discussed in chapter two, liberalisation had become the dominant ideological frame in Western Europe by the late 1980s. Most policy makers had come to view liberalisation as the best way to improve European competitiveness. Nevertheless, although most Member States may have agreed that some liberalisation was necessary, there was a significant divergence in the speed with which policy makers sought to liberalise and the methods used. The reform efforts that did occur in the late 1980s and 1990s were, to a large extent, coordinated by the Commission and structured within the institutional settings examined in this book.

Policy-making processes in other sectors are no doubt subject to different pressures and institutional opportunity structures from those found in the telecommunications sector. In the electricity sector, for example, the Commission was forced to rely on the EC's normal legislative procedures, rather than the treaty's competition rules, to achieve liberalisation (Schmidt 1998). In that sector, which was subject to less technological and economic pressure than telecommunications, the Treaty's competition rules did not provide the Commission with the increased agenda-setting authority that was uncovered in Part III. However, as Part III also pointed out, many factors influenced the ability of DG Competition officials to rely on their competition authority: these included normative considerations, the distribution of Member State preferences, and the coupling of a problem with its solution.

Exogenous technological and economic pressures may also have been more effective in forcing the liberalisation, as opposed to the harmonisation, of EC telecommunications. In fact, to a large extent, increased competition was inevitable. For example, technological developments had enabled users to bypass PTT network monopolies prior to the reform of most national regulatory frameworks. As such, exogenous pressures were a potent force for liberalisation. On the other hand, in areas such as licensing and numbering, most public and private actors argue that a continued role for national governments is necessary to foster a competitive market. The need for harmonisation of these national regulatory functions is somewhat less immediate. As the Member States maintained significant control over these areas, they were in a strong position to resist Commission harmonisation efforts. Thus, the asymmetrical impact of exogenous pressures may help to account for the ability of the Member States to hinder harmonisation in key areas (Part II) and also help to explain the Commission's success in coordinating full liberalisation (Part III).

Finally, some may contend that they are not convinced by the conclusion that the complexity of the EU policy-making process requires multiple approaches. Instead, it may be argued that this is merely a methodological point of view and that the opposite view could be maintained depending on the issue that is being examined and the goal of the project (for example, whether an attempt is being made to contribute to general knowledge or just to describe a particular case using scientific tools). Indeed, it is recognised that each research project has different aims and each policy area has different characteristics; whereas the telecommunications policy-making process is complex and dynamic, other policy areas may be less complicated and more stable, and developments may be more easily accounted for with a more limited theoretical repertoire. Nevertheless, to the extent that a researcher determines that more than one theoretical model may be useful

– or necessary – to explain policy development best in a particular sector or on a particular issue, a synthetic theoretical approach may help him or her better identify where – or at what stage of the policy process – a particular model may be most suitable.

10.3 TOWARDS A SYNTHETIC THEORETICAL APPROACH

Scholars grappling with the complex and multi-tiered dimensions of the EU policy-making process have sought to impose ordered understanding on processes which, while having some structure, have proved resistant to consistent analytical approaches and even agreed terminology. This book has demonstrated that no approach, on its own, captures the complexity of the policy-making process in the sector. Instead, insofar as they are useful at all, different approaches provide the best explanations for developments at different stages of the policy process.

Accordingly, this book employed a synthetic approach, which drew together a number of policy-making models, to analyse the development of a European telecommunications policy and to evaluate the functioning of the EU policy-making process. A single framework for analysis that combines multiple theoretical perspectives (Peterson and Bomberg 1999) is helpful in organising this analysis. The final section of the book seeks to contribute to the development of a theoretical model for EU policy making. A synthetic approach may be a valuable – perhaps necessary – tool for understanding and explaining EU policy development in many cases.

Super-systemic or Macro Level of Analysis

At the super-systemic or macro level of analysis, the overall development of the EU's telecommunications policy is conceptualised. The focus of this level of analysis is the process of integration or Europeanisation of the telecommunications sector and not the day-to-day legislative processes outlined in Parts II and III. At this level, the technological and economic pressures outlined in chapter two often drove the process. Developments at the super-systemic level are often best analysed through the use of meta theories.

Consistent with neofunctionalism, supranational institutions were important actors in the development of an EC telecommunications policy. Since the late 1970s, the Commission has served as a policy entrepreneur in the sector. By the early 1990s, the Commission had become an effective policy broker; it had marshalled political consensus in favour of reform;

had coordinated EC-wide liberalisation; and had institutionalised the role of private interests and external consultants at the EC's sub-systemic level. During the 1980s and 1990s, the ECJ supported Commission authority with weighty rulings and the EP gained a powerful voice over policy development.

At the same time, however, EC policy development was highly sensitive to national preferences, as Member State positions at the EC level were largely consistent with national interests. The UK, which is traditionally one of the strongest opponents of the transfer of competencies to the EC, became a strong advocate of increased Commission authority because its desired policy line was consistent with the Commission position (for example, the comitology dispute in chapters four and five and Commission usage of Article 86 in chapters seven and eight). Chapter six also found that the Member States were able to implement the Licensing Directive in ways that are consistent with the national ideologies, institutions and policy-making styles examined in chapter two. In fact, policy output on important issues generally reflected the preferences of the UK, French and German governments; in important areas, the Member States were able to hinder the transfer of authority to the EC.

Any attempt to explain the overall development of EC policy in the sector will thus probably need to rely on aspects of both neofunctionalist and intergovernmentalist theories. Kingdon's framework, which incorporates a more independent role for policy ideas, is also useful. This approach helps to analyse more precisely the timing and scope of institutional change, as the overall development of EC policy in the sector did not progress in a straight line. Indeed, there have been significant discontinuities as the political strength of the Commission was subjected to severe strains and setbacks. Kingdon's approach highlights the fact that the Commission's ability to act as a policy entrepreneur, in putting forth innovative policy proposals (a new policy frame) and affecting institutional change, was frequently dependent on the appearance of a policy window. Since ideas have played a vital role in the Commission's aspirations to Europeanise telecommunications, an approach that incorporates an independent role for ideas, such as Kingdon's approach, is particularly useful for understanding and explaining developments at the super-systemic level.

Systemic Level of Analysis

Analysis at the systemic level is focused on institutional politics. Negotiation between the EP, the Council and the Commission is the defining characteristic of the EC's systemic level of analysis. New institutionalism, specifically rational choice institutionalism, provides a useful starting point for analysis of the systemic level. The approach is well suited to analysing the constraints faced by a limited number of interest-maximising, strategic actors as they manoeuvre their way through a complicated set of EC Treaty rules and procedures.

The rules and procedures that govern the interaction of these institutional actors during this latter stage of the policy-making process are set out in the EC Treaty. Modification to the institutional setting is thus more difficult to achieve than in the sub-systemic level (where, as discussed below, the Commission has formal authority to initiate legislative proposals); institutional reform at the systemic level requires modification to the EC Treaty and suffers from the joint-decision trap (Scharpf, 1988) discussed in chapter six. Moreover, the systemic level institutional setting spans across sectors. Thus, the patterns of policy making uncovered in chapter five are likely to be similar to patterns found in other sectors. Rational choice institutionalism is thus also likely to be the most useful approach for explaining policy development at the systemic level of other policy sectors. At the same time, the framework nature of the EC Treaty rules probably leads to inter-institutional disputes over authority in sectors other than telecommunications.

The division of the policy-making process into levels of analysis is particularly helpful in areas of positive integration that rely on the EC's normal legislative procedures. In these areas, it is useful to distinguish between the two distinct stages of the EC policy-making process: the inter-institutional bargaining stage (or the systemic level of analysis) and the policy formulation stage (or the sub-systemic level of analysis, discussed further below). In focusing on institutional factors to distinguish levels of analysis, the line between the sub-systemic and systemic levels of analysis is clear. This contrasts with the problems discussed in the first chapter which the researcher encounters when attempting to apply Peterson's framework (1995a). A theoretical framework based entirely on institutional factors may lead to increased functionality and greater clarity in the boundaries of analysis. Future research should further explore the functionality of a systemic level of analysis defined, in a temporal sense, as the inter-institutional bargaining stage beginning when the Council and EP receive a draft proposal from the Commission.

In focusing on institutional politics, it is important to recognise the differences in the institutional settings for positive and negative integration. Although the Commission, the EP and the ECJ have each gained influence in a variety of sectors, formal agenda-setting authority is ultimately dependent upon the rules and procedures that govern policy development. In areas of positive integration, on the one hand, MEPs have a powerful right to veto EC legislation, the Council of Ministers has the final right of approval and the interaction between these institutions is laid out, to a large extent, in the EC Treaty. Nonetheless, the formal inter-institutional bargaining stage can last many months and there are sometimes disputes over authority.

On the other hand, in areas of negative integration, the Commission has the formal authority to bypass these potential difficulties. With the support of the ECJ, Commission officials have been able to issue directives unilaterally, while the formal approval of national representatives and MEPs is not required. Thus, in distinguishing between the rules and procedures that govern positive as opposed to negative integration, a great deal is learned about the relative importance of institutional actors.

The distinction between the two institutional environments is useful empirically. Chapter six demonstrated that the prevailing rules and procedures for positive integration enabled the Member States to construct a regulatory framework that left significant room for interpretation and diverse national implementation. Chapters seven and eight, on the other hand, demonstrated that the rules and procedures for negative integration enabled the Commission, with the support of the ECJ, to build a consensus in favour of its liberalisation programme. Although there were other factors that influenced policy development during this period, including technological and economic pressures and the hegemonic ideological frame of liberalisation and market competition, these pressures and influences were channelled through the different institutional settings. This led, in the first case, to a lack of harmonisation during the 1990s and, in the second case, to a coordinated liberalisation. Thus, the institutional focus helps to explain why certain reforms were easier to achieve than others.

The book also found that, in order to invoke its competition authority successfully in the pursuit of negative integration, the Commission must be particularly sensitive to the wider political climate. Within areas of negative integration, where the Commission maintains exclusive formal authority to draft and adopt legislation, the political climate plays a particularly large part in determining the Commission's approach to policy making, its relationship with other actors in the policy process, and its willingness to rely on its competition authority. For example, following Maastricht and the normative shift towards respecting subsidiarity in 1992, Commission officials retreated from an authoritative policy style and did not push their preferred policy line by resorting to the treaty's competition rules. A more accommodative policy style (Mazey and Richardson, 1997) emerged, thus prompting Commission officials to pursue widespread consultation and seek Member State agreement early in the process of policy formulation.

In other words, the Commission adjusted its policy style in accordance with the changed logic of appropriateness (March and Olsen, 1989). By the mid-1990s, as the policy environment became more favourable, the Commission consciously pursued a twin-track – or carrot and stick – strategy. It was still conscious of the need to take account of the wider political context (and thus continued to seek a political consensus behind its proposals), but nevertheless, recognising the more favourable political climate, also increasingly invoked its formal powers under Article 86 of the EC Treaty (formerly Article 90).

At the same time, the expansive definition of institutions under historical institutionalism is useful because norms are often a powerful component of the institutional environment. The normative shift towards respecting subsidiarity throughout the EC institutions during the early 1990s affected the policy-making process and policy outcomes. Rational choice institutionalism, on the other hand, better accounts for the strategic actions of institutional actors who jealously guard their authority and seek to influence institutional design. This approach is particularly useful at the systemic level, where the EC Treaty rules can be vague and contentious and can lead to inter-institutional disputes over authority.

Although a survey of institutional features goes a long way in analysing and explaining different stages of the policy-making process, institutions are not the only determinant of results. New institutionalism, in other words, cannot explain adequately the totality of the EU policy-making process. While an examination of the institutional framework serves as a good starting point for analysis, particularly at the systemic level, employing insights from actor-based approaches is necessary to gain a clearer picture of the influences on policy, particularly at the EC's sub-systemic level of analysis.

Sub-systemic Level of Analysis

This book defined the sub-systemic level of analysis, in a temporal sense, as the policy formulation stage until the Commission forwards a draft proposal to the Council of Ministers and the European Parliament. At the sub-systemic level, the EC Treaty gives the Commission the formal authority to initiate legislative proposals. This power gives the Commission the ability to set the agenda during the early stage of policy development. Every official EC legislative proposal is drafted, at least formally, by Commission officials and voted on by the full Commission before being forwarded to the other institutions for the inter-institutional bargaining stage.[1]

The Commission's formal authority to draft legislative proposals helps account for the focus that interest groups place on lobbying Commission officials. The large number of policy actors that seek to play a role in policy formulation also explains the utility of actor-based models in sub-systemic level analysis. In particular, the policy networks, epistemic communities and advocacy coalitions approaches were found to be useful in analysing and explaining the role of actors in policy formulation. At the same time, the structure of the policy formulation stage varies significantly from sector to sector; the institutional setting discussed in chapter four is specific to telecommunications. Due to the paucity of formal EC Treaty rules governing the policy formulation stage, policy-making is likely to be more fluid, and institutional modifications less cumbersome, than in the subsequent inter-institutional bargaining stage. Nevertheless, as policy formulation becomes increasingly institutionalised (Mazey and Richardson 2001), new institutionalism may become a more valuable tool for analysis of the EC's sub-systemic level.

Although the policy networks approach was applied to analyse the sub-systemic level, the epistemic community and advocacy coalition concepts frequently have more utility. The advocacy coalition concept should nevertheless be better defined as to the relationship between interests and beliefs in the formation of a coalition. Indeed, it may be easier and more useful in many cases to base advocacy coalitions on groups with common interests rather than groups with common belief systems. As Sabatier (1987, p. 683) acknowledges, 'the relative importance of "interests" versus "belief systems" needs to be explored. To what extent, for example, do economists' models based upon individual self-interest provide more parsimonious and equally valid explanations of policy change over time?' In short, how important is it to distinguish between an advocacy coalition based on shared interests as opposed to one based on shared belief systems? Sabatier has generally found it more difficult to identify a 'clear and falsifiable set of interests for most actors in policy conflicts' (p. 664).

In this case study, user interests in lower prices and more extensive and better services were clearly identifiable, as were TO interests in maintaining the *status quo*. Moreover, it may not be essential to distinguish between beliefs and interests in order to apply the approach usefully; a group of actors, made up of a variety of individuals and organisations, with shared or similar goals and demonstrating a coordinated approach over a period of time, would seemingly properly be termed an 'advocacy coalition', whether held together primarily by shared beliefs or shared interests. To the extent that the distinction between interests and beliefs is, however, found to be important and testable, it should be examined further.

10.4   CONCLUSION

In July 2003, the EU put into place an updated electronic communications framework that applies to all transmission networks in order to meet the internet-driven convergence of telecommunications, media and computers.

In this way, the EU has established one of the most sophisticated regulatory frameworks in the world. It highlights the Commission's success in Europeanising regulation of a sector that, for many years, was driven by the perceived need to catch up to the United States and Japan. The path to Commission leadership, however, was not straightforward; it was a complex interaction between many variables operating at different tiers of government and in different decisional arenas, and depended on the emergence of a transnational network of governmental and non-governmental institutions. There was a high degree of contingency about the steps that led to this development, and no single conceptual framework is able to account for it adequately.

To account for this complexity, this book thus relied on several policy-making theories to explain how and why the EU's telecommunications policy developed. It ventured to point towards a framework to help researchers determine the suitability of particular theoretical approaches to particular parts of the policy process. Although based on John Peterson's framework (1995b; Peterson and Bomberg 1999), the present case study was more firmly grounded in new institutionalism. Indeed, a survey of institutions (both at the systemic and sub-systemic levels of analysis, across various policy-making procedures and in different Member States) goes a long way towards explaining how policy developed in the sector.

In addition, other public policy concepts and frameworks were also needed to explain developments in this sector adequately. In particular, Kingdon's explanatory model of streams of politics, problems and ideas, provided valuable insight into the timing and scope of institutional and policy change. Actor-based models and concepts, including policy networks, epistemic community and advocacy coalition approaches, also helped to explain the interactions between institutions and numerous state and non-state actors in the policy process. These actor-based models were especially useful for examining the early stages of the policy-making process, when a wide range of actors, experts and interests sought to influence proposals. As such, a synthetic approach, which enables the researcher to apply a number of approaches to multiple settings and various stages or levels of analysis, is likely to be useful – even necessary – to understand and explain the many dimensions of EU policy-making in other areas.

The extent to which policy-making in telecommunications has been Europeanised is certainly remarkable. Notwithstanding various setbacks and difficulties faced by the Commission, it was able to coordinate the establishment of a comprehensive regulatory framework over telecommunications in little more than a decade after its initial consultation document, the 1987 Green Paper, was published. As a result, nearly every aspect of the sector has been liberalised and telecommunications operators in every Member State must now comply with the EU rules. Largely because of this, hundreds of millions of European telecommunications users now have access to higher quality, more advanced and less expensive telecommunications services. In coordinating the liberalisation and harmonisation of European telecommunications networks and services, the Commission has accomplished what would have been inconceivable to many people in the 1980s.

With the development of the 2003 electronic communications framework, the regulation of telecommunications was further Europeanised to the extent that all electronic networks and services, including fixed, mobile, broadcasting, satellite and internet networks and services, are now covered by the framework. Although the Commission gained veto authority over certain NRA decisions taken within the transnational European Regulators Group, it faced strong Member State resistance to its attempts to gain power to oversee implementation of the framework. Moreover, Member State opposition to further shifts of regulatory authority to the EU has undoubtedly been buttressed by the French and Danish rejections of the draft EU Constitution in May 2005. As a result, large-scale institutional change that would accompany the transfer to the EU of further, significant regulatory powers appears unlikely, at least for the foreseeable future. As a consequence, transnational regulatory networks, such as the European Regulators Group and the Communications Committee, will continue to be important forums for coordinating the implementation of EU telecommunications policies, as well as EU policies in other sectors.

NOTE

1. Of course, not all policy ideas are generated within the Commission; in fact, evidence suggests that a large majority of policy ideas are generated outside the Commission (Peterson and Bomberg 1999, p. 38).

Appendix 1.   Case study on the public consultation of the ONP Voice Telephony Directive

This appendix presents a case study on the changes made to the ONP Voice Telephony Directive as a result of the public consultation period. It is based on the section *Voting Patterns: ONP Voice Telephony Directive* in Marc Austin's March 1994 *Telecommunications Policy* article. The appendix, *Summary of comments on draft proposal for ONP Voice Telephony Directive, August–October 1991* was particularly useful. This analysis is also based on the original Commission documents, including the Commission's November 1991 draft proposal (and not the August 1992 draft as used by Austin), the Commission's own summaries of the submitted comments as found in the *Public Comments on the Analysis report on the application of ONP to voice telephony*, (Supplement to Document ONPCOM91-68, November 1991), and personal conversations with officials involved in the early stages of the policy formulation process. Six key issues are analysed.

SCOPE OF THE 'USERS' DEFINITION

Confirming that the policy formulation process was more favourable to users, the first issue examined (the scope of the term 'users' in the Analysis report) was resolved in favour of user interests. In particular, the definition of 'users' was modified in line with the comments expressed by user groups. Initially, the Analysis report did not include TOs in the definition of users. Annex 1 (Commission 1991b, p. 26) simply stated that '"users" means end-users and service providers.' Although no TOs commented on this omission during the consultation period, numerous user groups (including INTUG), service providers, business groups, and manufacturers' associations argued that TOs should be considered users when engaged in competitive offerings (Austin 1994, pp. 110–111). This advocacy coalition sought to establish the principle that competitive service providers should be served by the TOs on the same terms as the TOs' own competitive ventures (Commission

1991e). In line with their comments, Article 2 of the November 1991 proposal (Commission 1991c, p. 19) included in the definition of users, 'telecommunications organisations where the latter are engaged in providing services which are or may be provided also by others'. Thus, the policy formulation process favoured user interests, and did not merely reinforce TO interests, on at least one issue.

COMPULSORY COMPENSATION

The second issue examined (the requirement for compulsory compensation), was also resolved in favour of user interests, as the Analysis report was modified in line with the comments expressed by consumer organisations. Initially, the Analysis report (section 2.4) mandated that users have a right to compensation if the TO fails to meet the contracted service quality levels. Specifically, the TOs were required to provide compensation for delayed service provision and/or the non-availability of a service to which the user has subscribed (Commission 1991b, section 2.4.1, p. 3). The powerful consumer organisation, the *Bureau Européen des Unions de Consommateurs* (BEUC), along with another consumer organisation (CCC-Europe), supported this provision. Predictably, nearly all TOs firmly opposed the provision, arguing that they should not be liable for indirect or consequential losses, especially when multiple TOs are involved (Austin 1994, p. 111). Following the consultation period, the Commission not only retained the original wording, but strengthened the provision by adding into Article 6 of the November 1991 proposal (Commission 1991c, p. 21) that 'organizations representing consumer interests', in addition to the users themselves, can initiate legal proceedings against the TOs. This change is in line with the interests of the consumer organisations, which sought to enforce their role in the protection of European consumers. BEUC, in particular, was concerned specifically to protect small users, who may not have been able to initiate legal proceedings on their own (BEUC 1991). Thus, the first two issues examined were resolved in favour of users.

ADVANCED SERVICES PROVISION

The third issue examined (the provision of advanced services), however, was not resolved in favour of user interests. Instead, the Analysis report was modified in line with the comments expressed by the TOs. Although the Commission 'had hoped to establish mandatory timetables' for the TOs to provide advanced services (for example, call forwarding, freephone services, call transfer, charge reversal, directory services) (*CommunicationsWeek International*, p. 9), the Analysis report (sections 3.1 and 3.2) initially left the NRAs to 'ensure' their provision after taking into account the guideline dates set by the Commission, in consultation with the Communications Committee (Commission 1991b, p. 32). During the public comment period, nearly all the TOs adamantly opposed the establishment of a timetable, arguing that the provision of advanced services should be determined by market forces and not deadlines imposed by the Commission (Austin 1994, pp. 111–112). ETNO (1992) argued that the TOs themselves 'should have primary responsibility for the commercial decisions relating to their networks and services'. Following the public consultation period, the November 1991 proposal eliminated the role played by the Commission and the Communications Committee in setting guideline dates for the NRAs to follow. Instead, Articles 7 and 8 (Commission 1991c) required the NRAs to take 'into account [the] state of network development, market demand and progress with standardisation' when setting target dates. Although the provision for the establishment of a timetable was not completely eliminated, the changes are in line with the comments put forth by the TOs. Significantly, the resolution of this issue in favour of TO interests demonstrates that the (former) PTTs, although less powerful, were still important policy actors.

INTERCONNECTION AND SPECIAL NETWORK ACCESS

On the fourth issue examined (interconnection arrangements between competing networks), which was the most highly disputed issue during the public consultation period, the Analysis report was modified in line with comments put forth by both advocacy coalitions. Section 4.1 on *Interconnection and special network access* of the Analysis report mandated that the NRAs ensure that the TOs give network access to service providers and other TOs on request. The NRAs could intervene either to set fair, reasonable and non-discriminatory conditions or to ensure, in the interest of users, that agreements are implemented efficiently and in a timely manner and that they ensure the maintenance of end-to-end quality (Commission 1991b, p. 5). In line with historical institutionalism, in that past policy choices have a continuing effect on future policies, most TOs, along with ETNO, demanded that interconnection agreements be able to provide TOs with compensation for historical costs incurred in the construction of the public telecommunication infrastructures. Numerous service providers and user associations including INTUG, as well as the American Chamber of Commerce, the International Chamber of Commerce and the US Council for International Business put forth the counter-argument that 'equivalent' interconnection arrangements should be given to competitive service providers (Austin 1994, p. 112).

Following the public consultation period, the Commission put forth a provision that balanced the views expressed by the competing advocacy coalitions. In line with the interests of the TOs, the Commission revised the November 1991 proposal to include a provision in Article 9 allowing TOs to apply an access charge that could include reimbursement for historical costs. This provision was qualified, however, by the requirement that such charges 'be cost-oriented, non-discriminatory and fully justified, and must be levied only with the approval of the national regulatory authority acting in accordance with Community law' (Commission 1991c, p. 23). In line with the interests of users, the Commission introduced a provision (Article 9.2) obliging TOs that offer competitive services to provide non-discriminatory public network interconnection to competing service providers (Commission 1991c, p. 23). In striking a balance between the positions put forth during the public consultation, the Commission thus played the role of a 'broker' between TOs and users on this highly disputed issue.

TARIFF PRINCIPLES

On the fifth issue examined (tariff principles), the Analysis report was modified in line with the comments expressed by users. The Analysis report outlined principles that were designed to ensure that tariffs for the use of the public telephone network were transparent and cost-oriented. Section 5.1 mandated that tariffs should be independent of the information type and charges should 'normally' be itemised (Commission 1991b, pp. 7–8). During the public consultation period, the advocacy coalition of user groups, including ECTUA, consumer associations, service providers and manufacturer associations, along with external consultants such as Arthur Andersen and Arthur D. Little, supported the Commission proposal. The TOs, including ETNO, on the other hand, argued for a more flexible approach in which the TOs would be able to determine how tariffs are established (Austin 1994, pp. 112–113). In line with the comments of the users and external consultants, the Commission not only reproduced the tariff principles in Article 11 of the 1991 proposal, but included the additional provision that 'features additional to the provision of connection to the public telephone network and provision of voice telephone service shall, in accordance with Community law, be contracted separately and carry a separate tariff' (Commission 1991c, p. 25). This provision aimed to ensure an additional level of transparency in TO tariffing and was clearly in line with the interests of users.

COST-ACCOUNTING PRINCIPLES

On the final issue examined (cost-accounting principles), the Analysis report was also modified in line with the comments put forth by users. The Analysis report outlined a set of cost-accounting principles that were designed to ensure consistency and transparency in the accounting practices of the TOs. Section 5.2 noted that the cost accounting system was to be suitable for the enforcement of the tariff principles (as discussed above) (Commission 1991b, p. 8). Section 5.2 also detailed the costs that the system should identify (that is, direct, common and other costs) and how these costs were to be allocated. Accounting systems that did not conform to these requirements would have to be approved by both the NRA and the Commission (Commission 1991b, pp. 8–9).

Nearly all the organisations that submitted comments during the public consultation period put forth an opinion on this issue. Twelve TOs, along with ETNO, voiced opposition to this provision. They argued that the application of standard accounting procedures would be difficult and suggested that the NRAs be solely responsible for monitoring compliance. An advocacy coalition of 14 organisations, which included service providers, user groups, consumer organisations, manufacturer associations and business groups, demonstrated support for the principles outlined in section 5.2, and several proposed that an independent agency be established to monitor TO compliance (Austin 1994, p. 113). In line with the users once again, the Commission largely reproduced section 5.2 of the Analysis report in Article 12 of the 1991 proposal. The Commission even added a provision that mandated that the TOs give full financial account details to the NRA and the Commission on request (Commission 1991c, pp. 26–27). This provision aimed to ensure an additional level of transparency in TO accounting practices was clearly in line with the interests of users and adds further support to the argument that the policy formulation process no longer simply reinforces the interests of the former PTTs.

Thus, of the six issues examined, only one issue was resolved contrary to users interests, whereas four issues were resolved entirely in favour of users and one issue (interconnection) was resolved partly in line with user concerns. In short, the concerns of users, at the very least, were not entirely disregarded in the redrafting of the ONP voice telephony proposal.

Appendix 2.   Case study on the ONP Committee (now Communications Committee) meetings on the ONP Voice Telephony Directive

This appendix presents a case study on the changes made to the ONP Voice Telephony Directive as a result of two meetings of the ONP Committee (the Committee).[1] The changes made to the proposal by the Commission following the discussions with the Member States are evaluated by comparing the discussion in the Committee meetings on the November 1991 proposal with textual revisions made to the proposal forwarded to the Council and the EP in August 1992. The following analysis of four key voice telephony issues is based on the November 1991 proposal (Commission 1991c), the official minutes of the Committee meetings on 12–13 December 1991 (Commission 1991d) and 15–16 January 1992 (Commission 1992a), and the 1992 draft directive that was agreed to by the full Commission in July 1992 and forwarded to the Council and Parliament in August 1992 (Commission 1992b). Once again, it cannot be demonstrated with absolute certainty that individual comments made during the meetings led to specific textual revisions. The Commission officials who formulated the 1992 draft directive contend, however, and the draft itself (Commission 1992b, p. 9) acknowledges, that the proposal took 'into account ... discussion in the ONP Committee'. Therefore, the discussions at least influenced the redrafting of the proposal.

ADVANCED SERVICES PROVISION

During the Committee meetings, several of the national delegations expressed concern over the mandatory provision of advanced features (for example, call forwarding, freephone services, director services) in Articles 7 and 8 of the November 1991 proposal (Commission 1991d, pp. 4–5). The French and German delegations argued that some conditions in the November 1991 proposal were too demanding. In particular, these delegations felt that it would be too difficult and too costly to provide several of the advanced service features. They contended that their provision would not be possible on the analogue public telephone network, but only on an Integrated Services Digital Network (ISDN). On this issue, the arguments put forth by the French and German NRAs were in line with the interests of their national TOs (which would have been burdened with the provision of these advanced services). The Netherlands and the UK delegations, on the other hand, which had liberalised their domestic markets previously, generally supported the 1991 proposal and merely pointed to areas of the text that needed clarification.

In line with the comments expressed by the UK and Netherlands delegations, the Commission simplified and clarified the text. The Commission combined what had previously been two articles into one article (Article 8) in the 1992 draft directive (Commission 1992b, p. 28). Despite the comments expressed by the French and German delegations, the substance of the text and the advanced services to which Article 8 applies remained the same. On the issue of advanced services provision, therefore, the policy output from the new institutional setting was in line with the position taken by the more liberal Member States.

INTERCONNECTION AND SPECIAL NETWORK ACCESS

The conflict over an ONP interconnection policy continued into Committee discussions (Commission 1991d, pp. 5–6). Several delegations, including the UK, the Netherlands and Denmark, expressed concern that parts of the interconnection and special network access articles (9 and 10) of the November 1991 proposal (Commission 1991c, pp. 23–24) were unnecessary. For example, the Danish delegation pointed out that interconnection agreements were already being pursued in Denmark without ONP. Having already liberalised significantly, these Member States were concerned that a developing, comprehensive ONP interconnection regime would conflict with elements of their functioning, previously established national regulatory frameworks. Both France and Italy called for the elimination of Article 9.2, which had been added to the 1991 proposal following the public consultation period. Article 9.2 disadvantaged the national TOs in that it further developed the principle that both the TOs' own ventures and their competitors should face equal access conditions. The national positions of France and Italy can be explained, at least partly, by the close ties that their NRAs maintained with their national TOs. On the other hand, not having yet established a comprehensive national framework for interconnection, both the German and French delegations called for a detailed text on interconnection.

In line with historical institutionalism, the Commission's preference for a comprehensive interconnection policy led it to 'imprint its own image' (Hall 1986, p. 233) on the 1992 draft directive. This institutional bias was demonstrated by Ungerer's comment at the Committee meeting that 'without a policy on interconnection, ONP will be a joke!'[2] As early as the 1987 Green Paper, the Commission (1987, p. 157) had argued that an 'interconnection [policy] is urgently needed if distortion of trade and competition within the Community is to be avoided'. The Commission thus expanded the text on interconnection in line with the comments from the French and German delegations. 'Special network access' and 'interconnection' were separated into two more detailed Articles (9 and 10) (Commission 1992b, pp. 28–30). The Commission was also a 'policy broker' and made changes to the text in accordance with other comments of the national delegations. As the UK suggested, cross-border interworking was included in the interconnection article, while, as the French and Spanish delegations commented, the scope of the article was limited to 'reasonable' interconnection requests. Although the Commission did not respond to the UK and German request for clarification of the article's application to mobile telephony in the 1992 draft, these two Member States, through their increased institutional authority in the Council of Ministers, were able to get a detailed description of the article's application to mobile telephony included in the 1995 directive.

TARIFF PRINCIPLES

In the Committee discussions on tariff principles, the French and Italian delegations expressed concern that both tariff equalisation and the cross-subsidisation of business and residential users would not be allowed under Article 11 of the November 1991 proposal.[3] France and Italy were concerned that the provision of a universal service would not be possible under such conditions. The French tradition of telecommunications as a *service publique*, as examined in chapter two, continued to influence the position of French officials at the EC level. The UK and the Danish delegations, on the other hand, merely sought to clarify the text in some areas. In spite of the comments put forward, the Commission (1992b, p. 31) included essentially the same Article 11 in the 1992 draft directive. This demonstrates the institutional autonomy that the Commission maintains in the ONP policy formulation stage. In a subsequent Council of Ministers meeting over the draft proposal, however, France was able to 'logroll' and link its approval of an unrelated issue in order to establish tariff equalisation in the 1995 directive. The increased influence that the Member States maintain over ONP policy making in this different institutional context is examined further in chapters five and six.

COMMITTEE PROCEDURES (COMITOLOGY)

Several articles of the 1991 draft referred to a type IIIa regulatory committee to oversee the community-wide convergence and technical adaptation of the directive.[4] This committee gives the Commission, in consultation with the ONP Committee, the right to make binding legislation aimed at ensuring the proper implementation of the directive within the Member States. Under the type IIIa regulatory committee, the Member States can reject Commission proposals only through a qualified majority vote in favour of rejection. The French and German delegations, with the support of the Dutch, Irish, Spanish and Portuguese delegations, argued that some of the provisions subject to the committee procedure were too broad, and that some references to the committee should be replaced by reference to the normal legislative procedure for the approximation of internal market legislation (Commission 1992a, pp. 3–6). In putting forth these arguments, France and Germany sought to limit Commission influence over the harmonisation of their national regulatory frameworks. The UK, on the other hand, was the only delegation to favour a type IIIa regulatory committee. Having deregulated the UK telecommunications market well ahead of the other Member States, British officials favoured the type IIIa regulatory committee because they felt that the Commission was on their side.[5]

In the 1992 draft directive, the Commission changed the references to a committee procedure from a type IIIa regulatory committee to a type I advisory committee.[6] The advisory committee gives the Member States even less authority over implementation of the proposal. Docksey and Williams (1997, p. 134) note that it 'represents the most diluted form of Member State influence'. Under the advisory committee, the Commission maintains autonomous control, while the Member States can only put forth non-binding advice.

This change demonstrates both the Commission's autonomous ability to draft legislative texts in the policy formulation stage and its own preference for increased institutional authority. That is, the Commission reduced the ability of the Member States to inhibit reform (that is, delay implementation of ONP legislation) and increased its own power to achieve the re-regulation of telecommunications at the EC level (that is, ensure implementation of

ONP legislation). The rules and procedures governing the formulation of
ONP legislation (that is, the institutional design) have thus enabled the
Commission to become a key supranational actor in the harmonisation of
European telecommunications policies.

NOTES

1. As noted in Chapter 4, the ONP Committee was renamed the 'Communications Committee' in 2002.
2. Quoted in Austin (1994), p. 110.
3. Commission (1991d), pp. 6–7 discussions over Commission (1991c), Article 11, p. 25.
4. See Commission (1991c), arts. 22, 25, 27, 29.
5. Interview, UK Permanent Representative, Brussels, May 1999.
6. Commission (1992b), Articles 20, 23, 25, 27, 28, 29.

Bibliography

Adler, E. and Haas, P. (1992) 'Conclusion: Epistemic Communities, World Order, and the Creation of a Reflective Research Program', *International Organization* 46(1), Winter, 367–390.

Analysys Consultants (1989) *European Telecommunications 1: Standards and ONP – Keys to the Open Market*, Briefing Report Series, Cambridge: Analysys Publications.

Analysys Consultants (1991) *European Telecommunications 2: ONP, The Progress Report*, Briefing Report Series, Cambridge: Analysys Publications.

Analysys Consultants (1992) *ONP and After: A New Policy Agenda for Telecoms*, Briefing Report series, European Telecommunications 5, Cambridge: Analysis Publications.

Analysys Consultants (1994) *Network Europe: Telecoms Policy to 2000*, Briefing Report Series, European Telecommunications 9, Cambridge: Analysys Publications.

Armstrong, K. and Bulmer, S. (1998) *The Governance of the Single European Market*, Manchester: Manchester University Press.

Austin, M. (1994) 'Europe's ONP Bargain: What's in it for the User?' *Telecommunications Policy* 18(2): 97–113.

Benedetto, G. (2005) 'Rapporteurs as Legislative Entrepreneurs: the Dynamics of the Codecision Procedure in Europe's Parliament', *Journal of European Public Policy*, 12(1) February, 66–88.

BEUC (1991) *BEUC Position on the ONP Voice Telephony Directive*, Brussels: BEUC.

Bouwn, P. (2004) 'The Logic of Access to the European Parliament: Business Lobbying in the Committee on Economic and Monetary Affairs', *Journal of Common Market Studies*, 42(3), 473–495.

Bulmer, S. (1993) 'The Governance of the European Union: A New Institutionalist Approach', *Journal of Public Policy* 13(4): 351–380.

Bulmer, S. (1998) 'New Institutionalism and the Governance of the Single European Market', *Journal of European Public Policy* 5(3) September, 365–386.

Bulmer, S. and Scott, A. (1994) *Economic and Political Integration in Europe: Internal Dynamics and Global Context*, Oxford: Blackwell Publishers.

Burley, A.-M. and Mattli, W. (1993) 'Europe Before the Court: a Political Theory of Legal Integration', *International Organization*, 47(1), Winter, 41–76.

Burns, C. (2004) 'Codecision and the European Commission: A Study of Declining Influence?' *Journal of European Public Policy*, 11(1) February, 1–18.

Carpentier, Michel (1993) *European Community Telecommunications Policy*, speech at World Telecommunications Conference, London (8 December).

Caty, C.-F. and Ungerer, H. (1984) 'Les Télécommunications Nouvelle Frontière de l'Europe', *Futuribles*, 83, December, 29–50.

Cave, M. and Crowther, P. (1996) 'Determining the Level of Regulation in EU Telecommunciations: A Preliminary Assessment', *Telecommunications Policy*, 20(10): 725–738.

Cawson, A., Morgan, K., Webber, D., Holmes, P. and Stevens, A. (1990) *Hostile Brothers*, Oxford: Oxford University Press.

CEPT (1996) *'The Birth of Wide-ranging European Co-operation'*.

CEPT (2000) *'CEPT Response to the 1999 Communications Review'*, 14 February.

Cini, M. (1997) 'Administrative Culture in the European Commission', in Nugent, N. (ed.) *At the Heart of the Union: Studies of the European Commission*. London: Macmillan Press.

Cini, M. and McGowan, L. (1998) *Competition Policy in the European Union*. London: Macmillan Press.

Cohen, E. (1992) *Le Colbertisme 'High-Tech': Economie des Télécom et du Grand Projet*. Paris: Hachette.

Cohen, M., March, J.G. and Olsen, J.P. (1972) 'A Garbage Can Model of Organizational Choice', *Administrative Science Quarterly*, 17: 1–25.

Commission of the European Communities (1983) *Communication from the Commission to the Council on Telecommunications: Lines of Action*. COM (83) 573.

Commission of the European Communities (1984) *Communication from the Commission to the Council on Telecommunications: Lines of Action* COM (84) 277.

Commission of the European Communities (1985) *Completing the Internal Market: White Paper from the Commission to the European Council*, COM (85) 310 final.

Commission of the European Communities (1987) *Towards a Dynamic European Economy: Green Paper on the Development of the Common Market for Telecommunications Services and Equipment*, COM (87) 290 final.

Commission of the European Communities (1988a) *Towards a Competitive Community-Wide Telecommunications Market in 1992: Implementing the Green Paper on the Development of the Common Market for Telecommunications Services and Equipment*, COM (88) 48 final.

Commission of the European Communities (1988b) *Proposal for a Council Directive on the Establishment of the Internal Market for Telecommunications Services Open Network Provision*, COM (88) 825 final.

Commission of the European Communities (1988c) *Revised Proposal for a Council Directive on the Establishment of the Internal Market for Telecommunications Services through the Implementation of Open Network Provision*, COM (88) 325 final.

Commission of the European Communities (1990) *Commission Directive of 28 June 1990 on Competition in the Markets for Telecommunications Services* (90/388/EEC; OJ L192/10, 24.07.90).

Commission of the European Communities (1990a) *Open Network Provision (ONP) Consultation Procedures*, Open Network Provision Information Sheet 3, December.

Commission of the European Communities (1990b) *Explanatory Note: Re-examined Proposal for a Council Directive on the Establishment of the Internal Market for Telecommunications Services Through the Introduction of Open Network Provision* (ONP).

Commission of the European Communities (1991) *Telecommunications for Europe 1992: The CEC Sources, Volume 2*, Amsterdam: IOS Press.

Commission of the European Communities (1991a) *Guidelines on the Application of EEC Competition Rules in the Telecommunications Sector*, (91/C 233/02), OJ C 233/2.

Commission of the European Communities (1991b) Analysis Report on the Application of Open Network Provision to Voice Telephony, ONPCOM 91-53, 12 July 1991, OJ C 197, 26.7.91, p. 12.

Commission of the European Communities (1991c) *Draft Proposal for a Council Directive on the Application of Open Network Provision (ONP) to Voice Telephony*, ONPCOM91-67, 22 November.

Commission of the European Communities (1991d) *Minutes of the 12th Meeting of the ONP Committee held on 12/13 December 1991*, ONPCOM 91-80.

Commission of the European Communities (1991e) *Public Comments on the Analysis Report on the Application of ONP to Voice Telephony*, Supplement to Document ONPCOM91-68, November.

Commission of the European Communities (1991f) *Implementation of the ONP Framework Directive: 1991 Work Programme for the Development of ONP Conditions*, Open Network Provision Information Sheet 4, March, p. 2.

Commission of the European Communities (1992a) *Minutes of the 13th Meeting of the ONP Committee Held on 15/16 January 1992*, ONPCOM 92-05.

Commission of the European Communities (1992b) *Proposal for a Council Directive on the Application of Open Network Provision (ONP) to Voice Telephony*, COM (92) 247 final- SYN 437, 27 August.

Commission of the European Communities (1992c) *Review of the Situation in the Telecommunications Services Sector*, SEC (92) 1048, 21 October.

Commission of the European Communities (1993) *RACE 1993*. Brussels: CEC.

Commission of the European Communities (1993a) *Commission Communication on the Consultation on the Review of the Situation in the Telecommunications Services Sector*, COM(93) 159 Final, 28 April 1993.

Commission of the European Communities (1993b) *White Paper on Growth, Competitiveness, and Employment: The Challenges and Ways Forward into the 21st Century*, COM (93) 700 final, Brussels, 5 December.

Commission of the European Communities (1994) *Twenty-Third Report on Competition Policy 1993*, Luxembourg: CEC.

Commission of the European Communities (1994a) *Communication on the Consultation on the Green Paper on Mobile and Personal Communications*, COM(94) 492 final, 23 November 1994.

Commission of the European Communities (1995) *RACE 1995*. Brussels: CEC.

Commission of the European Communities (1996) *Proposal for a European Parliament and Council Directive on a Common Framework for General Authorizations and Individual Licenses in the Field of Telecommunications Services*, COM (95) 545 final (96/C 90/05) – 95/0282(COD), Submitted by the Commission on 30 January.

Commission of the European Communities (1997) *Directive of the European Parliament and of the Council on a Common Framework for General Authorisations and Individual Licenses in the Field of Telecommunications Services*, COM 97(17), 10 April, OJ L 117, 17 May.

Commission of the European Communities (1999) *The 1999 Communications Review*, COM (1999) 539, November.

Commission of the European Communities (1999a) *Fifth Report on the Implementation of the Telecommunications Regulatory Package*, COM (1999) 537 final, Brussels, 10 November 1999.

Commission of the European Communities (2000) *Draft Directive on the Authorisation of Electronic Communications Networks and Services*, COM (2000) 386, 12 July.

Commission of the European Communities (2000a) *Public Consultation Results of 1999 Communications Review*, COM (2000) 239, 26 April.

Council of the European Communities (1984a) *Council Recommendation of 12 November 1984 Concerning the Implementation of Harmonization in the Field of Telecommunications*, (84/549/EEC; OJ L298/49, 16.11.94).

Council of the European Communities (1984b) *Council Recommendation of 12 November 1984 Concerning the First Phase of Opening Up Access to Public Telecommunications Contracts*, (84/550/EEC; OJ L298/51, 16.11.94).

Council of the European Communities (1985) *Council Decision of 25 July 1985 on a Definition Phase for a Community Action in the Field of Telecommunications Technologies: R&D Programme in Advanced Communication Technologies for Europe (RACE)*, (85/372/EEC; OJ L 210/24, 07.08.85).

Council of the European Communities (1987) *Council Decision of 14 December 1987 on a Community Programme in the Field of Telecommunications Technologies: Research and Development (R&D) in Advanced Communications Technologies in Europe (RACE Programme)*, (88/28/EEC; OJ L 16/35, 21.01.88).

Council of the European Communities (1988) *Council Resolution of 30 June 1988 on the Development of the Common Market for Telecommunications Services and Equipment up to 1992*, (88/C 257/01; OJ C257/1, 04.10.88).

Council of the European Communities (1990) *Council Directive of 28 June 1990 on the Establishment of the Internal Market for Telecommunications Services through the Implementation of Open Network Provision*, (90/387/EEC, OJ L192/1, 24.07.90).

Council of the European Communities (1991) *Council Directive of 29 April 1991 on the Approximation of the Laws of the Member States concerning Telecommunications Terminal Equipment, Including the Mutual Recognition of their Conformity*, (91/263/EEC), OJ L 128/1.

Council of the European Communities (1992) *Council Directive of 11 May 1992 on the Adoption of Standards for Satellite Broadcasting of Television Signals*, (92/38/EEC).

Council of the European Communities (1993) *Council Directive of 29 October 1993 Supplementing Directive 91/263/EEC in Respect of Satellite Earth Station Equipment*, (93/97/EEC).

Cox, G.W. (1999) 'The Empirical Content of Rational Choice Theory, A Reply to Green and Shapiro', *Journal of Theoretical Politics*, 11: 147–169.

Craig, P. and De Búrca, G. (1995) *EC Law: Text, Cases, and Materials*, Oxford: Clarendon Press.

Crombez, C. (2001) 'The Treaty of Amsterdam and the co-decision procedure', in Schneider, G. and Aspinwall, M. (eds) *The Rules of Integration: Institutionalist Approaches to the Study of Europe*. Manchester: Manchester University Press, pp. 101–122.

Dandelot, M. (1993) *Le Secteur des Télécommunications en France: Rapport au Ministre de l'Industrie, des Postes et Télécommunications et du Commerce Extérieur*, Paris: PTT Ministry, July.

Dang-Nguyen, G. (1985) 'Telecommunications: a Challenge to the Old Order', in Sharp, M. (ed.) *Europe and the New Technologies*, London: Frances Pinter.

Dang-Nguyen, G. (1988) 'Telecommunications in France', in Foreman-Peck, J. and Müller, J. (eds) *European Telecommunications Organisation*, Baden-Baden: Nomos, pp. 131–154.

Dawkins, W. (1987) 'Industry Outlines Telecoms Policy for EEC', *Financial Times*, 23 January, p. 3.

Dehousse, R. (1998) *The European Court of Justice: The Politics of Judicial Integration*, London: Macmillan Press.

Delcourt, B. (1991) 'EC Decisions and Directives on Information Technology and Telecommunications', *Telecommunications Policy*, 14: 15–21.

Department of Trade and Industry (1983), *The Future of Telecommunications: Government Policy Explained*, London: HMSO.

DG Information Society (2000), *The Authorisation of Electronic Communications Networks and Services*, Working Document, 27 April.

Dimitrakopoulos, D. (2001) 'Learning and Steering: Changing Implementation Patterns and the Greek Central Government', *Journal of European Public Policy*, 8(4): 604–622.

Docksey, C. and Williams, K. (1997) 'The Commission and the Execution of Community Policy', in Edwards, G. and Spence, D. (eds), *The European Commission*, London: Cartermill International.

Dolowitz, D. (2000) *Policy Transfer and British Social Policy: Learning from the USA?* Buckingham: Open University Press.

Downs, A. (1967) *Inside Bureaucracy*, Boston: Little Brown & Co.

Dyson, K. (1980) *The State Tradition in Western Europe*, Oxford: Martin Robertson.

Dyson, K. (1982) 'West Germany: The Search for a Rationalist Consensus', in Richardson, J. (ed.), *Policy Styles in Western Europe*, London: Allen & Unwin, pp. 17–46.

Dyson, K. and Humphreys, P. (eds) (1986) *The Politics of the Communications Revolution in Western Europe*, London: Frank Cass.

Dyson, K. and Humphreys, P. (eds) (1990) *The Political Economy of Communications: International and European Dimensions*, London: Routledge.

Earnshaw, D. and Judge, D. (1993) 'The European Parliament and the Sweeteners Directive', *Journal of Common Market Studies*, 31(1): 103–116.

Earnshaw, D. and Judge, D. (1996) 'From Co-operation to Co-decision: The European Parliament's Path to Legislative Power', in Richardson, J. (ed.) *European Union: Power and Policy-Making*, London: Routledge, pp. 96–126.

Eberlein, B. and Grande, E. (2005) 'Beyond Delegation: Transnational Regulatory Regimes and the EU Regulatory State', *Journal of European Public Policy*, 12(1): 89–112.

Eberlein, B. and Kerwer, D. (2004) 'New Governance in the European Union: A Theoretical Perspective', *Journal of Common Market Studies*, 42(1): 121–142.

ECTUA (1987) ECTUA *Position on European Telecommunications Policy*, March.

Edwards, G. and Spence, D. (eds) (1997) *The European Commission*, London: Cartermill International.

Egeberg, M., Schaefer, G. and Trondal, J. (2003) 'The Many Faces of EU Committee Governance', *West European Politics*, 26(3): 19–40.

Egyedi, T.M. (1993) *Eurocentrism in International Standardisation: The Forces Driving Change*, Contribution to the ITS European Regional Conference, 20–22 June.

Eliassen, K. and Sjovaag, M. (eds) (1999) *European Telecommunications Liberalisation*, London: Routledge.

Elliot, C. and Schlaepfer, R. (2001) 'The Advocacy Coalition Framework: Application to the Policy Process for the Development of Forest Certification in Sweden', *Journal of European Public Policy*, 8(4): 642–661.

Esser, J. and Noppe, R. (1996) 'Private Muddling Through as a Political Programme? The Role of the European Commission in the Telecommunications Sector in the Eighties', *West European Politics*, 19(3): 547–562.

Ester, J. (1989) *The Cement of Society: A Study of Social Order*, Cambridge: Cambridge University Press.

ETNO (1992) *Comments on ONP Applied to Voice Telephony*, Common Position CP08.

ETO (1998) *Regulating Operators With Significant Market Power*, study produced for the Commission, Copenhagen: ETO.

ETO (1999a) *Categories of Authorisations*, study prepared for the Commission, Copenhagen: ETO.

ETO (1999b) *Information Required for Verification*, study prepared for the Commission, Copenhagen: ETO.

ETO (1999c) *Fees for Licensing Telecommunications Services and Networks*, study prepared for the Commission, Copenhagen: ETO.

ETO (1999d) *The Numbering Requirements of Corporate Telecommunications Networks (CN) and Their Impact on Public Network Numbering*, study prepared for the Commission, Copenhagen: ETO.

European Parliament (1989) *Legislative Resolution on the Proposal for a Council Directive on the Establishment of the Internal Market for Telecommunications Services Open Network Provision*, (Doc. Ad-122/89) 26 March, OJ C 158/296.

European Parliament (1990) *Draft Recommendation of the EMAC Concerning the Common Position of the Council with a View to Adopting a Directive on the Establishment of the Internal Market for Telecommunications Services Through the Implementation of Open Network Provision*, (Doc. C3–36/90) 29 March, PE 139.188.

European Parliament (1993) *Legislative Resolution on the Proposal for a Council Directive on the Application of Open Network Provision to Voice Telephony*, (Doc. A3–0064/93) 26 April, OJ Č 115/105.

European Parliament (1995) *Legislative Resolution on the Proposal for a Council Directive on the application of Open Network Provision to Voice Telephony*, (Doc. A4–0231/95) 20 November, OJ C 308/112.

European Parliament (1999) *Activity Report 1 November 1993 – 30 April 1999 on the Codecision Procedure*, PE 230.998, May.

Flynn, G. (1995) *Remaking the Hexagon: the New France in the New Europe*, Boulder: Westview Press.

Foreman-Peck, J. and Müller, J. (eds) (1988) *European Telecommunications Organisation*, Baden-Baden: Nomos.

From, J. (2002) 'Decision-Making in a Complex Environment: A Sociological Institutionalist Analysis of Competition Policy Decision-Making in the European Commission', *Journal of European Public Policy*, 9(2): 219–237.

Fuchs, G. (1995) 'The European Commission as Corporate Actor? European Telecommunications Policy After Maastricht', in Rhodes, C. and Mazey, S. (1995), *The State of the European Union: Building a European Policy*, Vol. 3, Boulder, CO: Lynne Rienner, pp. 413–429.

Gains, F., Peter, J. and Stoker, G. (2005) 'Path Dependency and the Reform of English Local Government', *Public Administration*, 83(1): 25–45.

GAP (1988) *Open Network Provision (ONP) in the Community*, Report, Brussels: Commission of the European Communities, 20 January.

Garfinkel, L. (1994) 'The Transition to Competition in Telecommunication Services', *Telecommunications Policy*, 18(6): 427–431.

Garrett, G. (1995) 'The Politics of Legal Integration in the European Union', *International Organization* 49(1): 171–181.

Gillick, D. (1991) 'Telecommunications Policy in the UK: Myths and Realities', *Telecommunications Policy*, 15(1): 3–9.

Gillick, D. (1992) 'Telecommunications Policies and Regulatory Structures: New Issues and Trends', *Telecommunications Policy*, 16: 726–732.

Gorges, M. (2001) 'The New Institutionalism and the Study of the European Union: The Case of the Social Dialogue', *West European Politics*, 24(4): 152–168.

Grant, W. (1993) 'Pressure Groups and the EC', in Mazey, S. and Richardson, J. (eds) (1993) *Lobbying in the European Community*, Oxford: Oxford University Press.

Green Cowles, M. (1995) 'Setting the Agenda for a New Europe: the ERT and EC 1992', *Journal of Common Market Studies*, 33(3): 127–147.

Greenwood, J. (1997) *Representing Interests in the European Union*, London: Macmillan Press.

Greenwood, J., Grote, J.R. and Ronit, K. (eds) (1992) *Organised Interests and the European Community*, London: Sage.

Grossman, E. (2004) 'Bringing Politics Back In: Rethinking the Role of Economic Interest Groups in European Integration', *Journal of European Public Policy* 11(4): 637–654.

Haas, E. (1968) *The Uniting of Europe: Political, Social, and Economic Forces, 1950–7*, Stanford: Stanford University Press.

Haas, P. (1992) 'Introduction: Epistemic Communities and International Policy Co-ordination', *International Organization*, 46(1): 1–35.

Haas, P. (2004) 'When Does Power Listen to Truth? A Constructivist Approach to the Policy Process', *Journal of European Public Policy*, 11(4): 569–592.

Haid, A. and Müller, J. (1988) 'Telecommunications in the Federal Republic of Germany,' in Foreman-Peck, J. and Müller, J. (eds) *European Telecommunications Organisation*, Baden-Baden: Nomos.

Hall, P. (1986) *Governing the Economy: The Politics of State Intervention in Britain and France*, Cambridge: Polity Press.

Hall, P. (1992) 'The Movement from Keynesianism to Monetarism: Institutional Analysis and British Economic Policy in the 1970s', in Steinmo, S. *et al.* (eds) *Structuring Politics: Historical Institutionalism in Comparative Analysis*, Cambridge: Cambridge University Press, pp. 90–113.

Hall, P. (1994) 'Pluralism and Pressure Politics', in Hall, P. *et al.* (eds) *Developments in French Politics*, London: Macmillan.

Hall, P. Hayward, J. and Machin, H. (eds) (1994) *Developments in French Politics*, London: Macmillan.

Hall, P. and Taylor, R. (1996) 'Political Science and the Three New Institutionalisms', *Political Studies*, XLIV: 936–957.

Hawkins, R. (1993) 'Changing Expectations: Voluntary Standards and the Regulation of European Telecommunications', *Communications and Strategies*, September.

Hayward, J.E.S. (1986) *The State and the Market Economy: Industrial Patriotism and Economic Intervention in France*, Brighton: Wheatsheaf Books.

Heclo, H. (1978) 'Issue Networks and the Executive Establishment', in King, A. (ed.) *The New American Political System*, Washington, DC: AEI.

Heclo, H. and Wildavsky, A. (1974) *The Private Government of Public Money*, London: Macmillan.

Héritier, A. (1996) 'The Accommodation of Diversity in European Policy-Making and its Outcomes: Regulatory Policy as a Patchwork', *Journal of European Public Policy*, 3(2): 149–167.

High Level Group on the Information Society (1994) *Europe and the Global Information Society: Recommendations to the European Council* The Bangemann Report, http://europa.eu.int/ISPO/infosoc/backg/bangemann.html (last visited 4 October 2005).

Hildebrand, P.M. (1992) 'The European Community's Environmental Policy, 1957 to 1992: From Incidental Measures to an International Regime?' *Environmental Politics*, 1: 13–44.

Hills, J. (1986) *Deregulating Telecoms: Competition and Control in the United States, Japan and Britain*, Westport, CT: Quorum Books.

Howorth, J. (2004) 'Discourse, Ideas, and Epistemic Communities in European Security and Defence Policy', *West European Politics*, 27(2): 211–234.

Hull, R. (1993) 'Lobbying Brussels: A View from Within', in Mazey, S. and Richardson, J. (eds), *Lobbying in the European Community*, Oxford: Oxford University Press.

Humphreys, P. (1986) 'Legitimating the Communications Revolution: Governments, Parties and Trade Unions in Britain, France and West Germany', in Dyson, K. and Humphreys, P. (eds) *The Politics of the Communications Revolution in Western Europe*, London: Frank Cass, pp. 165–194

Immergut, E. (1992) 'The Rules of the Game: The Logic of Health Policy-making in France, Switzerland, and Sweden', in Steinmo, K., Thelen, K. and Longstreth, F. (eds) *Structuring Politics: Historical Institutionalism in Comparative Analysis*, Cambridge: Cambridge University Press.

INTUG (1989) *Telecommunications Initiatives Related to the Single Market in Europe: Background Status and Action Steps*, Brussels: INTUG.

INTUG (1993) *Response to the Services Directive Review*, January, Brussels: INTUG.

Kassim, H. (1994) 'Policy Networks, Networks and European Union Policy Making: A Sceptical View', *West European Politics*, 17(4): 15–27.

Kelemen, D. (2002) 'The Politics of 'Eurocratic' Structure and the New European Agencies', *West European Politics*, 25(4): 93–118.

Kenis, P. and Schneider, V. (1991) 'Policy Networks and Policy Analysis: Scrutinizing a New Analytical Toolbox', in Marin, B. and Mayntz, R. (eds) *Policy Networks: Empirical Evidence and Theoretical Considerations*, Frankfurt: Westview Press, pp. 25–59.

Keohane, R.O. (1984) *After Hegemony: Cooperation and Discord in the World Political Economy*, Princeton, NJ: Princeton University Press.

Kiessling, T. and Blondeel, Y. (1998) 'The EU Regulatory Framework in Telecommunications: A Critical Analysis', *Telecommunications Policy*, 22(7): 571–592.

King, M. (2005) 'Epistemic Communities and the Diffusion of Ideas: Central Bank Reform in the United Kingdom', *West European Politics*, 28(1): 94–123.

Kingdon, J. (1995) *Agendas, Alternatives, and Public Policies*, New York: Harper Collins.

Krasner, S.A. (1984) 'Approaches to the State: Alternative Conceptions and Historical Dynamics', *Comparative Politics*, 16: 223–246.

Krasner, S.A. (1989) 'Sovereignty: An Institutional Perspective', in Caporaso, J.A. (ed.) *The Elusive State: International and Comparative Perspectives*, Newbury Park, CA: Sage.

Kübler, D. (2001) 'Understanding Policy Change with the Advocacy Coalition Framework: An Application to Swiss Drug Policy', *Journal of European Public Policy*, 8(4): 623–641.

Labarrère, Cf. C. (1985) *L'Europe des Postes et des Télécommunications*, Paris: Masson.

Levy, D. (1999) *Europe's Digital Revolution: Broadcasting Regulations, the EU and the Nation State*, London: Routledge.

Lodge, J. (1994) 'Transparency and Democratic Legitimacy', *Journal of Common Market Studies*, 32(3): 343–368.

Lord, C. (2003) 'The European Parliament in the Economic Governance of the European Union', *Journal of Common Market Studies*, 41(2): 249–267.

Lowndes, V. (1996) 'Varieties of New Institutionalism: A Critical Appraisal', *Public Administration*, 74, Summer, 181–197.

Mansell, R. (1993) 'Latecomers or Innovators? The European Policy Challenge', in Mansell, R. *The New Telecommunications*, London: Sage Publications.

March, J. and Olsen, J. (1984) 'The New Institutionalism: Organizational Factors in Political Life', *The American Political Science Review*, 78: 734–749.

March, J. and Olsen, J. (1989) *Rediscovering Institutions*, New York: Free Press.

Marin, B. and Mayntz, R. (eds) (1991) *Policy Networks: Empirical Evidence and Theoretical Considerations*, Frankfurt: Westview Press.

Marsh, David (1991) 'Privatization Under Mrs. Thatcher: A Review of the Literature', *Public Administration*, 69, Winter, 459–480.

Marsh, D. and Rhodes, R.A.W. (eds) (1992) *Policy Networks in British Government*, Oxford: Clarendon Press.

Maurer, A. (2003) 'The Legislative Powers and Impact of the European Parliament', *Journal of Common Market Studies*, 41(2): 227–247.

Mayntz R. and Schneider V. (1988) 'The Dynamics of System Development in a Comparative Perspective: Interactive Videotex in Germany, France and Britain', in R. Mayntz and T.P. Hughes (eds), *The Development of Large Technical Systems*, Frankfurt, New York: Campus, pp. 263–298.

Mazey, S. (1995) 'The Development of EU Equality Policies: Bureaucratic Expansion on Behalf of Women?', *Public Administration*, 73: 591–609.

Mazey, S. (2001) 'European Integration: Unfinished Journey or Journey Without an End?' in Richardson, J. (ed.) (2001) *European Union: Power and Policy-Making*, London, Routledge.

Mazey, S. and Richardson, J. (eds) (1993) *Lobbying in the European Community*, Oxford: Oxford University Press.

Mazey, S. and Richardson, J. (1994) *Promiscuous Policy-Making: The European Policy Style?* European Public Policy Institutute, Occasional Papers, no. 94/2, Coventry: EPPI.

Mazey, S. and Richardson, J. (1996) 'The logic of Organisation: Interest Groups', in Richardson, J. (ed.) *European Union: Power and Policy-Making*, London: Routledge, pp. 200–215.

Mazey, S. and Richardson, J. (1997) 'The Commission and the Lobby', in Edwards, G. and Spence, D. (eds) *The European Commission*, London: Cartermill, pp. 178–212.

Mazey, S. and Richardson, J. (2001) 'Institutionalizing Promiscuity: Commission-Interest Group Relations in the EU', in Stone Sweet, A., Sandholtz, W. and Fligstein, N. (eds), *The Institutionalization of Europe*, Oxford: Oxford University Press, pp. 71–93.

Mazey, S. and Richardson, J. (2001a) 'The Commission and the Lobby: Reconciling Openness and Effective Policy-Making', in Edwards, G. and Spence, D. (eds) *The European Commission*, London: Cartermill.

McKendrick, G.G. (1987) 'The INTUG View on the EEC Green Paper', *Telecommunications Policy*, 11: 325–329.

Menon, A. and Kassim, H. (eds) (1996) *The European Union and National Industrial Policy*, London: Routledge.

Monar, J. (1994) 'Interinstitutional Agreements: The Phenomenon and its New Dynamics after Maastricht', *Common Market Law Review*, 31: 693–719.

Moon, J., Richardson, J.J. and Smart, P. (1986) 'The Privatisation of British Telecom: A Case Study of the Extended Process of Legislation', *European Journal of Political Research*, 14: 339–355.

Moravcsik, A. (1994) 'Preferences and Power in the European Community: A Liberal Intergovernmentalist Approach', in Bulmer, S. and Scott, A. (eds) *Economic and Political Integration in Europe: Internal Dynamics and Global Context*, Oxford: Blackwell Publishers.

Morgan, K. and Webber, D. (1986) 'Divergent Paths: Political Strategies for Telecommunications in Britain, France, and West Germany', in Dyson, K. and Humphreys, P. (eds) *The Politics of the Communications Revolution in Western Europe*, London: Frank Cass.

NERA (1997) *Issues Associated with the Creation of a European Regulatory Authority for Telecommunications*, study prepared for the Commission, London, April.

Nora, S. and Minc, A. (1978) 'L'Informatisation de la Société', Report to the President of the Republic, Paris: Seuil.

North, D.R. (1990) *Institutions, Institutional Change, and Economic Performance*, Cambridge: Cambridge University Press.

Nugent, N. (1994) *The Government and Politics of the European Union*, London: Macmillan Press.

Nugent, N. (ed.) (1997) *At the Heart of the Union: Studies of the European Commission*, London: Macmillan Press.

OFTEL (1986) *Annual Report for 1985*, London: OFTEL.

OFTEL (1996) *Pan European Regulator: Oftel's Views*, London: OFTEL.

Osborne, Iain (2000) *Telecoms: Liberalisation without Harmonisation?* The Federal Trust: European Essay No. 6. London: Federal Trust.

Paoloni, Marc (1987) 'Les P et T contre l'Europe?' *La Tribune*, 24 January.

Pelkman, S.J. and Young, D. (1998) *Telecoms-98*, Brussels: Centre for European Policy Studies.

Peters, B.G. (1994) 'Agenda-Setting in the European Community', *Journal of European Public Policy*, 1(1): 9–26.

Peters, B.G. (1996) 'Agenda-Setting in the European Community', in Richardson, J. (ed.) *European Union: Power and Policy-Making*, London: Routledge, pp. 61–76.

Peters, B.G. (1999) *Institutional Theory in Political Science: The 'New Institutionalism'*, London: Pinter.

Peterson, J. (1992) 'The European Technology Community: Policy Networks in a Supranational Setting', in Marsh, D. and Rhodes, R.A.W. (eds) *Policy Networks in British Government*, Oxford: Clarendon Press, pp. 226–248.

Peterson, J. (1995a) 'Decision-Making in the European Union: Towards a Framework for Analysis', *Journal of European Public Policy*, 2(1): 69–93.

Peterson, J. (1995b) 'Policy Networks and European Union Policy Making: A Reply to Kassim', *West European Politics*, 18(2): 389–407.

Peterson, J. (2001) 'The Choice for EU Theorists: Establishing a Common Framework for Analysis', *European Journal of Political Research*, 39: 289–318.

Peterson, J. and Bomberg, E. (1999) *Decision-Making in the European Union*, London: Macmillan.

Peterson, J. and Sharp, M. (1998) *Technology Policy in the European Union*, London: Macmillan.

Pierson, P. (1993) 'When Effect Becomes Cause: Policy Feedback and Political Change', *World Politics*, 45: 595–628.

Pierson, P. (1996) 'Path to European Integration', *Comparative Political Studies*, 29(2): 123–163.

Pierson, P. (2000) 'Increasing Returns, Path Dependence and the Study of Politics', *American Political Science Review*, 94(2): 262.

Pollack, M. (1996) 'The New Institutionalism and EC Governance: The Promise and Limits of Institutional Analysis', *Governance: An International Journal of Policy and Administration*, 9(4): 429–458.

Pollack, M. (1997) 'The Commission as an Agent', in Nugent, N. (ed.) *At the Heart of the Union: Studies of the European Commission*, London: Macmillan Press, pp. 109–128.

Pollack, M. (1997a) 'Delegation, Agency and Agenda-Setting in the European Union', *International Organization*, 51(1): 99–134.

Pollack, M. (1998) 'The Engines of Integration? Supranational Autonomy and Influence in the European Union', in Sandholtz, W. and Sweet, A.S. (eds) *European Integration and Supranational Governance*, Oxford: Oxford University Press.

Prévot, H. (1989) *Le Débat Public: Rapport de Synthese*, Paris: Ministere des Postes, des Telecommunications et de L'Espace, August, pp. 163–182.

Putnam, R.D. (1988) 'Diplomacy and Domestic Politics: The Logic of Two-Level Games', *International Organization*, 42(3): 427–460.

Radaelli, C.M. (1995) 'The Role of Knowledge in the Policy Process', *Journal of European Public Policy*, 2(2): 159–183.

Rein, H. and Schön, D. (1991) 'Frame-Reflective Policy Discourse', in Wagner, P. and Weiss, C.H. (eds) *Social Sciences, Modern States: National*

Experiences and Theoretical Crossroad, Cambridge: Cambridge University Press, pp. 262–289.

Rhodes, C. and Mazey, S. (1995) *The State of the European Union: Building a European Policy, Vol. 3*, Boulder, Colorado: Lynne Rienner Publishers.

Rhodes, M. (2000) 'Recasting European Welfare States', *West European Politics*, vol. 23(2), special edition.

Rhodes, R.A.W. (1997) *Understanding Governance: Policy Networks, Governance, Reflexivity and Accountability*, Oxford: Oxford University Press.

Rhodes R.A.W., Bache I. and George S. (1996) 'Policy Networks and Policy-Making in the European Union: A Critical Appraisal', in Hooghe, L. (ed.) *Cohesion Policy and European Integration,* Oxford: Oxford University Press, pp. 367–387.

Rhodes R.A.W. and Marsh, D. (eds) (1993) *Policy Networks in British Government*, Oxford: Oxford University Press.

Richardson, J. (ed.) (1982) *Policy Styles in Western Europe*, London: Allen & Unwin.

Richardson, J. (1996) 'Eroding EU Policies: Implementation Gaps, Cheating and Re-steering', in Richardson, J. (ed.) *European Union: Power and Policy-Making*, London: Routledge, pp. 278–294.

Richardson, J. (1996a) 'Policy-Making in the EU: Interests, Ideas and Garbage Cans of Primeval Soup', in Richardson, J. (ed.) *European Union: Power and Policy-Making*, London: Routledge, pp. 3–24.

Richardson, J. (ed.) (1996b) *European Union: Power and Policy-Making*, London: Routledge.

Richardson, J. (ed.) (2001) *European Union: Power and Policy-Making*, 2nd ed., London: Routledge.

Richardson, J. and Jordan, G. (1979) *Governing Under Pressure: The Policy Process in a Post-Parliamentary Democracy*, Oxford: Martin Robertson.

Rose, R. (1989) *Politics in England: Change and Persistence*, London: Macmillan.

Roundtable of European Industrialists (1986) *Clearing the Lines: A User's View on Business Communications in Europe*, Paris and Brussels: European Roundtable Secretariat.

Sabatier, P. (1987) 'Knowledge, Policy-Oriented Learning, and Policy Change: An Advocacy Coalition Framework', *Knowledge: Creation, Diffusion, Utilization*, 8(4): 649–692.

Sait, E.M. (1938), *Political Institutions: A Preface*, New York: D. Appleton-Century.

Sandholtz, W. (1992a). 'ESPRIT and the Politics of International Collective Action', *Journal of Common Market Studies*, 30: 1–22.

Sandholtz, W. (1992b) *High-Tech Europe: The Politics of International Cooperation*, Berkeley and Los Angeles: University of California Press.

Sandholtz, W. (1996) 'Membership Matters: Limits of the Functional Approach to European Institutions', *Journal of Common Market Studies*, 34: 403–429.

Sandholtz, W. (1998) 'The Emergence of a Supranational Telecommunications Regime', in Sandholtz, W. and Sweet, A.S. (eds) *European Integration and Supranational Governance*, Oxford: Oxford University Press, pp. 134–163.

Sandholtz, W. and Sweet, A.S. (eds) (1998) *European Integration and Supranational Governance*, Oxford: Oxford University Press.

Scharpf, F. (1986) 'Policy Failure and Institutional Reform: Why Should Form Follow Function?' *International Social Science Journal*, 108: 179–181.

Scharpf, F. (1988) 'Joint Decision Trap: Lessons from German Federalism and European Integration', *Public Administration*, 66: 239–278.

Scharpf, F. (1991) 'Games Real Actors Could Play: The Challenge of Complexity', *Journal of Theoretical Politics*, 3: 277–304.

Scharpf, F. (1997) *Games Real Actors Play: Actor-Centered Institutionalism in Policy Research*, Boulder, Colorado: Westview Press.

Scharpf, F. (1999) *Governing in Europe: Effective and Democratic?* Oxford: Oxford University Press.

Schiller, D. (1982) *Telematics and Government*, Norwood, NJ: Ablex.

Schmidt, S. (1991) 'Taking the Long Road to Liberalization: Telecommunications Reform in the Federal Republic of Germany', *Telecommunications Policy*, June, 209–222.

Schmidt, S. (1996a) 'Sterile Debates and Dubious Generalisations: European Integration Theory Tested by Telecommunications and Electricity', *Journal of Public Policy*, 16(3): 233–271.

Schmidt, S. (1996b) 'Privatizing the Federal Postal and Telecommunications Services', in Benz, A. and Goetz, K.H. (eds), *A New German Public Sector? Reform, Adaptation and Stability*, Dartmouth: Aldershot, pp. 45–70.

Schmidt, S. (1998) 'Commission Activism: Subsuming Telecommunications and Electricity under European Competition Law', *Journal of European Public Policy*, 5(1): 169–184.

Schmidt, V. (1996) *From State to Market? The Transformation of French Business and Government*, Cambridge: Cambridge University Press.

Schneider, G. and Aspinwall, M. (eds) (2001) *The Rules of Integration: Institutionalist Approaches to the Study of Europe*, Manchester: Manchester University Press.

Schneider, V. (1992) 'Organised Interests in the European Telecommunications Sector', in Greenwood, J. *et al*, (eds) *Organised Interests and the European Community*, London: Sage.

Schneider, V., Dang-Nguyen, G. and Werle, R. (1994) 'Corporate Actor Networks in European Policy-Making: Harmonizing Telecommunications Policy', *Journal of Common Market Studies*, 32: 473–498.

Schneider, V. and Werle, R. (1990) 'International Regime or Corporate Actor? The European Community in the Telecommunications Policy', in Dyson, K. and Humphreys, P. (eds) *The Political Economy of Communications: International and European Dimensions*, London: Routledge, pp. 77–106.

Schneider, V. and Werle, R. (1991) 'Policy Networks in the German Telecommunications Domain', in Marin, B. and Mayntz, R. (eds) *Policy Networks: Empirical Evidence and Theoretical Considerations*, Frankfurt: Westview Press, pp. 97–136.

Scott, W.R. and Meyer, J.W. (eds) (1994) *Institutional Environments and Organizations*, Thousand Oaks, CA: Sage, Chapters 11 and 12.

Solomon, J. (1984) 'The Future Role of International Telecommunications Institutions', in *Telecommunications Policy*, 8: 213–222.

Solomon, J. (1986) 'Telecommunications Evolution in the UK', *Telecommunications Policy*, September.

Steinfield, C., Bauer, J. and Caby, L. (1995) *Telecommunications in Transition: Policies, Services, and Technologies in the European Community*, London: Sage.

Steinmo, K., Thelen, K. and Longstreth, F. (eds) (1992) *Structuring Politics: Historical Institutionalism in Comparative Analysis*, Cambridge: Cambridge University Press.

Steunenberg, B. (1994) 'Decision Making under Different Institutional Arrangements: Legislation by the European Community', *Journal of Institutional and Theoretical Economics*, 150: 642–669.

Stone Sweet, A. and Brunell, T. (1998) 'The European Court and the National Courts: A Statistical Analysis of Preliminary References 1961–95', *Journal of European Public Policy*, 5(1): 66–97.

Stone Sweet, A., Sandholtz, W. and Fligstein, N. (eds) (2001) *The Institutionalization of Europe*, Oxford: Oxford University Press.

Suleiman, E. (1978) *Elites in French Society*, Princeton: Princeton University Press.

Thatcher, M. (1995) 'Regulatory Reform and Internationalization in Telecommunications', in Hayward, J.E.S. (ed.) (1995) *Industrial Enterprise and European Integration*, Oxford: Oxford University Press, pp. 239–269.

Thatcher, M. (1996) 'The European Community and High Technology: The Importance of the National and International Context', in Menon, A. and Kassim, H. (eds), *The European Union and National Industrial Policy*, London: Routledge, pp. 178–203.

Thatcher, M. (1997a) 'The Development of European Regulatory Frameworks: The Expansion of European Community Policy-Making in Telecommunications', in Stravridis, E. *et al.* (eds) *New Challenges to the European Union*, Aldershot: Ashgate.

Thatcher (1997b) 'L'Impact de la Communauté Européenne sur la Réglementation Nationale: Les Services Publics en France et en Grande-Bretagne', *Revue Politiques et Management Public*, 15(3): 141–168.

Thatcher, M. (1999a) *The Politics of Telecommunications: National Institutions, Convergence, and Change in Britain and France*, Oxford: Oxford University Press.

Thatcher, M. (1999b) *The Europeanisation of Regulation: The Case of Telecommunications*, EUI Working Paper RSC No. 99/22, Florence: European University Institute.

Thatcher, M. (1999c) 'Liberalisation in Britain: From Monopoly to Regulation of Competition', in Eliassen, K. and Sjovaag, M. (eds), *European Telecommunications Liberalisation*, London: Routledge.

Thatcher, M. (2001) 'The Commission and National Governments as Partners: EC Regulatory Expansion in Telecommunications 1979–2000', *Journal of European Public Policy*, 8(4): 558–584.

Thatcher, M. (2004) 'Winners and Losers in Europeanisation: Reforming the National Regulation of Telecommunications', *West European Politics*, 27(2): 284–309.

Thelen, K. and Steinmo, S. (1992) 'Historical Institutionalism in Comparative Politics', in Steinmo, K., Thelen, K. and Longstreth, F. (eds) *Structuring Politics: Historical Institutionalism in Comparative Analysis*, Cambridge: Cambridge University Press.

Thimm, A. (1992) *America's Stake in European Telecommunications Policies*, Westport, CT: Quorum Books.

Tsebelis, G. (1990) *Nested Games: Rational Choice in Comparative Politics*, London: University of California Press, p. 96.

Tsebelis, G. (1994) 'The Power of the European Parliament as a Conditional Agenda Setter', *American Political Science Review*, 88(1): 128–142.

Ungerer H. and Costello N. (1989) *Telecommunications in Europe: Free choice for the User in Europe's 1992 Market*, Luxembourg: Commission of the European Communities.

Vickers, J. and Yarrow, G. (1986) 'Telecommunications: Liberalisation and the Privatisation of British Telecom', in Kay, J. *et al.* (eds) *Privatisation and Regulation: the UK Experience*, Oxford: Clarendon Press.

Watkinson, J. and Guettler, J. (1999) *The Possible Added Value of an ERA for Telecommunications*, study prepared for the European Commission, November, London: Eurostrategies/ Cullen International.

Weaver, R.K. and Rockman, B.A. (eds) (1993) *Do Institutions Matter?* Washington, D.C.: The Brookings Institution.

Weiler, J. (1982) 'Community, Member States and European Integration: Is the Law Relevant?' *Journal of Common Market Studies*, 20: 399–356.

Weiler, J. (1997) 'The Reformation of European Contitutionalism', *Journal of Common Market Studies*, 35(1): 97–131.

Weiss, Ernst (1980) 'Focus on Europe, User Group Survey: Deutsche Telecom eV, West Germany', *Telecommunications Policy*, September, 228–230.

Wellens, K. and Borchardt, G. (1989) 'Soft Law in European Community Law', *European Law Review*, 14(4): 267–321.

Werle, R. (1999) 'Liberalisation of Telecommunications in Germany', in Eliassen, K. and Sjovaag, M. (eds), *European Telecommunications Liberalisation*, London: Routledge, pp. 110–127.

Westlake, M. (1994) *A Modern Guide to the European Parliament*, London: Pinter.

Wheeler, J. (1992) 'Key Issues in Europe's Open Network Provision: The Case of the German VANS Providers', *Telecommunications Policy*, January/February, 85.

Wilson, F. (1987) *Interest Group Politics in France*, Cambridge: Cambridge University Press.

Wincott, D. (1995) 'The Role of Law or the Rule of the Court of Justice?' *Journal of European Public Policy*, 2(4): 603–623.

Wincott, D. (1996) 'The Court of Justice and the European Policy Process', in Richardson, J. (ed.), *European Union: Power and Policy-Making*, London: Routledge, pp. 170–184.

Worthy, J. and Kariyawasam, R. (1998) 'A Pan-European Telecommunications Regulator?' *Telecommunications Policy*, 22(1): 1–7.

Wright, V. (1997) 'Introduction: La Fin du *Dirigisme?*' *Modern and Contemporary France*, 5(2): 151.

Xavier, P. (1998) 'The Licensing of Telecommunications Suppliers: Beyond the EU's Directive', *Telecommunications Policy*, 22(6): 483–492.

Zito, A. (2001) 'Epistemic Communities, Collective Entrepreneurship and European Integration', *Journal of European Public Policy*, 8(4): 585–603.

Index

access conditions 26, 78, 116, 173, 233
Alcatel 35, 188
Amsterdam Treaty (1997) 112, 121
Analysys study 76–7, 78, 79, 80, 89,
    172, 175–6, 178
anti-dumping duties 28
AOL Europe 226
APEC 190, 201
Arthur D. Little study 59, 172
AT&T 28, 29
ATLAS 39
Austin, Marc 78, 84, 92, 94, 96
Austria
    cable network 199, 203
    infrastructure liberalisation 198
    liberalisation 191, 198, 206
    mobile communications 206
    NERA study 136
    veto powers 207, 228

Baker, Kenneth 31
Bangemann, Martin 118, 121, 134,
    174, 188, 197
Bangemann Report 188–9, 246
BBC 226
Beesley, Professor Michael 31
Belgium
    cable network 199, 203
    compensation, compulsory 120
    competition 161, 162
    ECTUA 84
    liberalisation 174, 176, 178, 206
    licensing 143, 151
    mobile communications 206
    ONP Directive 110, 113, 118
    Services Directive 168
    terminal equipment regulation 164,
        165
    veto powers 207
BEUC 85, 201
Birmingham declaration (1992) 177

Boetsch, Wolfgang 198–9
Borrell, Jose 136, 206
British Telecom case 63–4, 162, 168
Brittan, Leon 163, 166, 170, 171, 172,
    174
broadband 36, 61, 62–3, 163, 164
broadcasting 18, 226, 232, 237
Bulgaria 236

Cable Directive 199, 200, 201, 202–3,
    234
cable network, Germany 199, 200, 202,
    203
cable services 28
    APEC 190, 201
    France 36, 199, 202, 203
    Green Paper 200–202
    infrastructure liberalisation 199–200,
        202, 203
    United States 28–9, 200
'call-me-back' equipment 26–7
Carpentier, Michel 26, 81, 134
Carsberg, Sir Bryan 33
CEPT 50–55, 63, 148–9, 192, 218–19,
    234, 242–4
    Commission, relationship with 57,
        61–2, 66, 84–5, 133, 217, 220,
        221–2
    and licensing 5, 133, 141, 142–3,
        144–9, 221–2, 237
    SSA 62
    see also SOGT
CERP (postal committee) 145, 146
Channel Plus 188
Chirac, Jacques 30, 32, 37, 38, 66
CNCL 38, 39
College of Commissioners 193
comitology 96, 113, 115, 118–20, 123,
    126, 250
Comitology Decision (1987) 121
Commissioners' independence 193–4